CONTENTS

iv

CHAPTER 1

Introduction

1.1 GETTING ACQUAINTED WITH MICROSOFT EXCEL

Microsoft Excel is a software package known as a "spreadsheet," but it can also be thought of simply as a place to organize things and make calculations. When you open up Excel on your computer, you will see a bunch of little rectangles, called "cells." In each of these cells, you can enter just about anything you want. And you can format each cell anyway you want, and resize it to any size you want, and also do just about any calculation that you want. Excel is really an extremely powerful tool. Let's look at some of the things you can do with it.

1.2 ENTERING DATA

One thing that you will be particularly interested in doing as you study statistics is entering data into Excel so that you can then organize it and analyze it. All you have to do is click on the cell where you want to put the first entry and type it in. If you want to put the next data value into the *same* row, then hit TAB to go to the next cell in that row and type in the next value. If you want to put the next data value into the *next* row, then hit ENTER to go to the next cell in that column. (If you want to put all of the data into a single row, hit TAB between entries, and if you want to put all of the data into a single column, hit ENTER between entries. If you want the data to be in rows *and* columns, then hit TAB until you get to the end of each row and hit ENTER to go to the beginning of the next row.) You can also use the arrow keys to go from one cell to another. Try typing in a bunch of numbers into various cells in an Excel spreadsheet in order to see how it works.

When you enter data, it is often a good idea to type in labels for the data as well. For example, a table showing the 2001 sales of seven U.S. companies has been typed into Excel with column headings and formatting in the figure below. See if you can copy it!

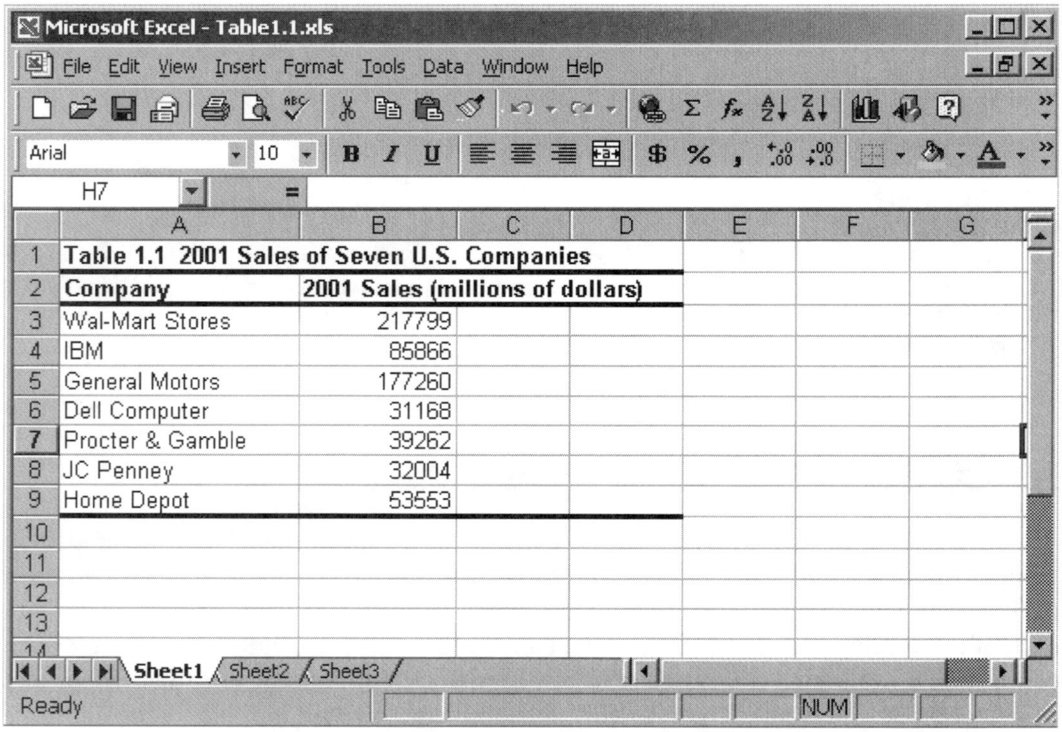

Table 1.1 2001 Sales of Seven U.S. Companies

Company	2001 Sales (millions of dollars)
Wal-Mart Stores	217799
IBM	85866
General Motors	177260
Dell Computer	31168
Procter & Gamble	39262
JC Penney	32004
Home Depot	53553

Figure 1.1 A table of data entered into a Microsoft Excel spreadsheet (with formatting).

When you enter numbers to be used in calculations, do not put commas in them. When you start entering data in the second column (column B), hitting ENTER will bring you to the next cell *in that column*.

1.3 FORMATTING CELLS

To get things like the dark borders that you see in the table above, you have to format the cells. Highlight the cells that you want to format, and go to **Format>Cells…** and then click on the **Border** tab in the window that pops up.

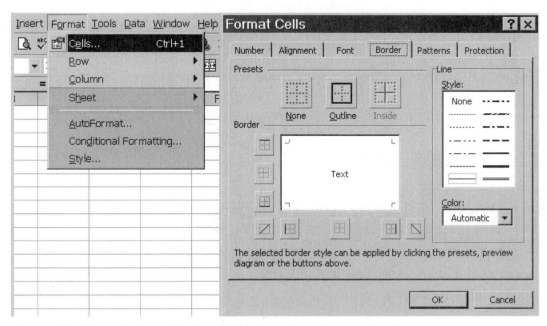

Figure 1.2 Menus for formatting the border of a cell or cells.

Click on the style of border that you want (the thick, heavy line was used above) and then where you want it to appear on the cell (top and bottom were selected above). Then click **OK.**

Other tabs in the **Format>Cells…** dialog box allow you to format the cell based on its contents (number vs. text, currency, percentage, etc.), change the color of the cell, change the font style or color, or change the way text appears in the cell. There are so many options that you just have to "click around" and try them in order to see what they do and when you might want to use them.

In order to resize cells (to make the columns wider so that you can see all of the text inside them, for example), simply click on the border between the lettered column headings and drag it over as far as you wish. The **A** and the **B** columns were widened for the table above.

Another, sometimes simpler and quicker and better, method for resizing columns so that you can see everything in them is to highlight the columns, then go to **Format>Column>AutoFit Selection**. Then the columns will all be exactly as wide as needed!

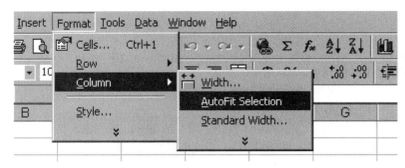

Figure 1.3 Resizing columns so that they are exactly as wide as needed.

1.4 BASIC CALCULATIONS

To perform any mathematical calculation, click on the cell where you want the answer to appear. To begin any calculation, type the equals symbol: =. Then type in the calculation that you're interested in. If you want to use a value from a specific cell, you can just click on that cell instead of typing the value.

For example, type in "=2+3+4" (without the quotes) and hit ENTER. You'll see 9 appear! But then if you click on the cell again, you'll see what you had originally typed in the formula bar above the cells.

A1	▼	=	=2+3+4				
	A	B	C	Formula Bar	E	F	G	H
1	9							
2								
3								
4								

Figure 1.4 Viewing a calculation's result and formula at the same time.

Excel works like a scientific calculator, performing operations in the correct order. Type in "=2+3*4" (without the quotes) and you will see 14, because it multiplies before it adds. There are also several built-in functions. If you want the square root of 49, for example, you can type "=SQRT(49)" and you'll see 7 appear. For exponentiation, use the "^" symbol. To see all of the built-in functions available to you, go to **Insert>Function** (or click on the f_x icon on the toolbar).

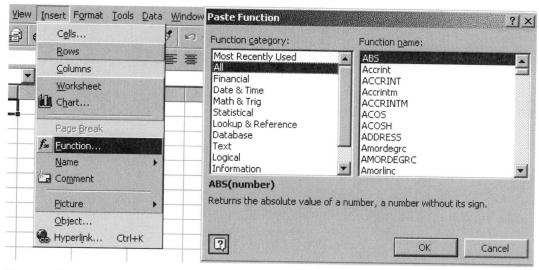

Figure 1.5 Menus for inserting a built-in function.

Excel keeps track of which functions you've most recently used, so they are listed there in case you need to use them again. Click on **All** to see the entire list of functions (in alphabetical order). You might recognize some of the functions in the **Math & Trig** category. For this course, you'll mainly be using the ones in the **Statistical** category.

1.5 CELL REFERENCES

When performing calculations, you can either enter *numbers* into formulas or just references to cells that *contain* the numbers. For example, you can type 49 into cell A1, hit ENTER to go to cell A2, and type "=SQRT(" (without the quotes), click on cell A1 (which automatically types "A1" after the open parenthesis), type ")" and hit ENTER. You'll see 7 appear! When you click on cell A2 and look up at the Formula Bar, you'll see "=SQRT(A1)." Change the 49 in cell A1 to a 36 and you'll see the 7 below it change to a 6! This can be very useful.

If you have a whole set of repetitive calculations to make, Excel's relative cell references are very helpful. For example, suppose you want to find the square roots of all of the numbers, 1-50. Enter 1 into cell A1 and 2 into cell A2. Continue on until you have 50 in cell A50. (Actually, you can highlight cells A1 and A2 and copy it down to cell A50 by dragging a plus sign in the bottom right corner!) Now, in cell B1, type "=SQRT(" (without the quotes), click on cell A1, and hit ENTER. Copy cell B1 and paste it into cells B2-B50 (or drag it down by its bottom right corner). The reference to cell A1 changes to which*ever* cell is just to the left, and you should see all of the numbers' square roots.

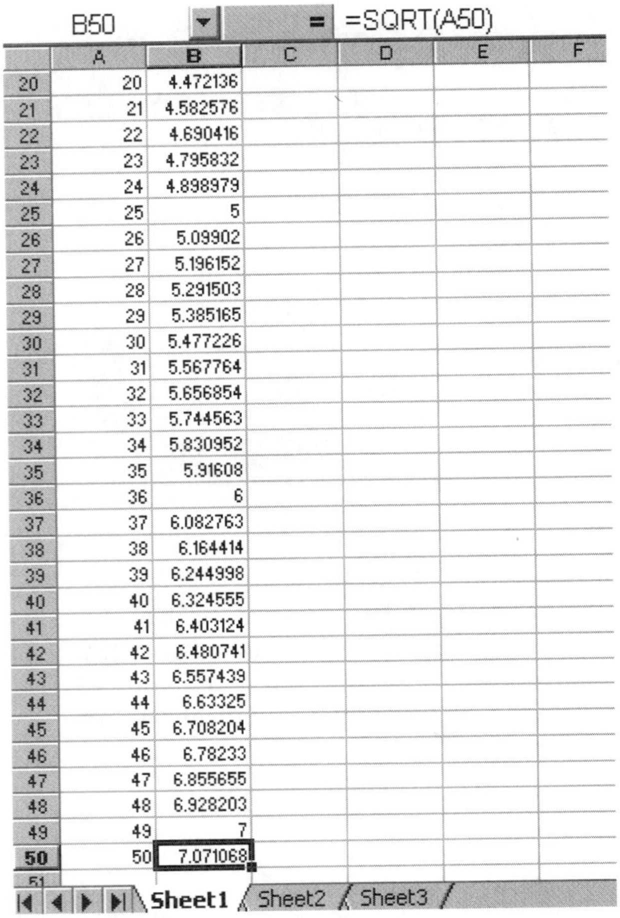

B50 = =SQRT(A50)

	A	B	C	D	E	F
20	20	4.472136				
21	21	4.582576				
22	22	4.690416				
23	23	4.795832				
24	24	4.898979				
25	25	5				
26	26	5.09902				
27	27	5.196152				
28	28	5.291503				
29	29	5.385165				
30	30	5.477226				
31	31	5.567764				
32	32	5.656854				
33	33	5.744563				
34	34	5.830952				
35	35	5.91608				
36	36	6				
37	37	6.082763				
38	38	6.164414				
39	39	6.244998				
40	40	6.324555				
41	41	6.403124				
42	42	6.480741				
43	43	6.557439				
44	44	6.63325				
45	45	6.708204				
46	46	6.78233				
47	47	6.855655				
48	48	6.928203				
49	49	7				
50	50	7.071068				

Figure 1.6 Square roots of the numbers up to 50, found by using Excel's copying feature and relative cell references.

If you do *not* want the cell reference to change when you are pasting a function to a new location, then you need to make it "absolute" by typing a $ before the column and/or row reference (whichever one or ones you want to keep the same). In the square root example above, you would want it to read "=SQRT($A$1)." Then pasting it to a new location (or copying it down) would copy *exactly* the same thing (and exactly the same reference).

1.6 WORKSHEETS

When you open a new (blank) Excel workbook, you will see three sheet tabs at the bottom of the window.

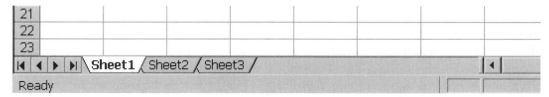

Figure 1.7 Worksheet tabs at the bottom of the window.

There are several worksheets available in a single Excel file (called a workbook) so that you can organize your work with some on one worksheet and some on another. You can also rename a worksheet by right-clicking on the sheet tab and selecting **Rename**. If you no longer need a sheet, you can delete it by going to the **Edit** menu and selecting **Delete Sheet**.

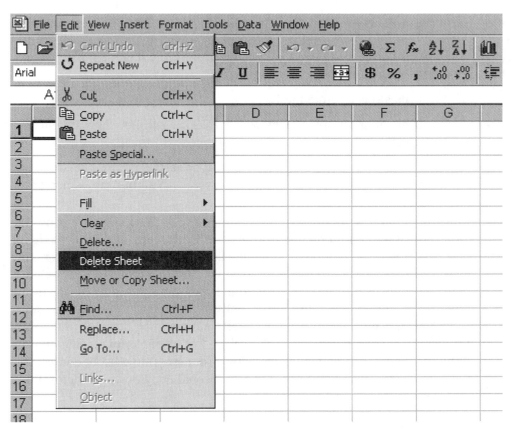

Figure 1.8 The **Edit** menu, where options for deleting, moving, or copying worksheets are located.

If you want to have two worksheets that contain at least some of the same information, you can copy a worksheet into the same workbook (or into another workbook) by selecting the next option on the **Edit** menu shown above. This option can also be used for moving sheets, but it's easier simply to drag the sheet tab at the bottom to the desired new location instead. (For example, you could click on **Sheet2** and drag it to the left of **Sheet1** so that it would actually be the first sheet.) And finally, you can always add a new sheet (go to **Insert>Worksheet**) at any time. When there are too many sheets to see all of their tabs at once, you have to use the arrows just to the left of the sheet tabs in order to scroll through the sheets. (The arrows to the right are for scrolling through the single current sheet that is showing.)

1.7 TIPS AND TRICKS

The more you use Excel, the more you'll get used to it and probably enjoy it. There are a few "tips and tricks" that you may be interested in, though, so here they are.

> Excel automatically formats something like "1-5" as "5-Jan," interpreting it as a date. This can be frustrating if you really meant it to be a range of data values from 1 to 5! What you need to do is tell Excel that this is really text, not a date. There are 3 ways you can do this:

  1. Type a space before the 1.

  2. Type an apostrophe: ' before the 1. This will not show up but tells Excel that this cell should have a simple text format.

  3. Format the cell as text before you type in the 1-5 by clicking on it, going to **Format>Cell...** and click on the **Number** tab. Then select **Text.**

> There are several ways to copy and paste in Excel. You can always use the copy and paste icons on the toolbar, or go to **Edit>Copy** and **Edit>Paste** (or **Edit>Paste Special**), but don't forget the keyboard strokes: CTRL-C for copy and CTRL-V for paste. If you want to copy a cell to a group of cells below it or to the right of it, click on the cell and then drag the little square in the bottom right corner (while the cursor looks like a + sign) to the right or down as far as you want. Highlighting two cells and dragging down or to the right will copy whatever *pattern* is in the cells. For example, two cells with the numbers 1 and 2 in them would copy down (or across) to a list of consecutive numbers. Two cells with 2 and 4 in them would copy down (or across) to a list of consecutive *even* numbers. Try it!

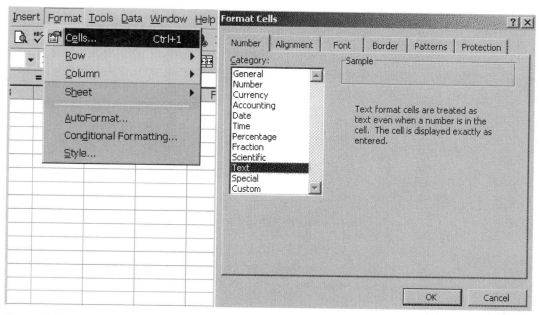

Figure 1.9 Formatting a cell to text only. (Just be sure that you realize that text-formatted cells can NEVER be used in calculations, even if they contain numbers! They can be used for labels on charts, though.)

➢ There are three options for editing a cell without losing its current contents. 1) Double-click on it. 2) Click on it and hit F2. 3) Click on it and then click on the Formula Bar above where you see its entire contents. Then you can edit as you wish.

➢ A shortcut for making cell references partially absolute or fully absolute is to highlight the reference in the Formula Bar and then press F4 until you have as many $'s as you want. (This will toggle through all of the options: A1, the relative reference, $A$1, the absolute reference, A$1, which will keep the row number the same, and $A1, which will keep the column the same.) The dollar signs make it so the column letter or row number won't change when it's copied somewhere else.

➢ If you are filling out a text box in one of Excel's function windows or wizard windows and need to see what's in the cells behind it, you can either just click on it and drag it out of the way, or you can click on the collapse icon: to the right of the text box. Then click on the expand icon: in order to get the window back again.

➢ Don't forget Excel's Help menu! Any time you have a question on how to do something, take advantage of it. Just go to **Help>Microsoft Excel Help** (or hit the F1 key). You can search by entering a keyword (using the **Index**) or by simply typing in a question (using the **Answer Wizard**) to get a list of related options. There is a wealth of information there!

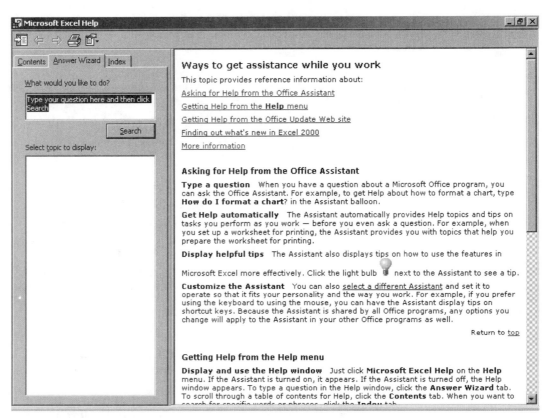

Figure 1.10 Microsoft Excel's Help menu.

Exercises

1.1. Enter the following data into an Excel Spreadsheet, sort it, and format it to currency.

$25,000 $23,750 $20,200 $15,900 $32,250 $35,600 $42,100 $29,400

1.2. Enter the following data into an Excel Spreadsheet, sort it, and format it to percentages.

0.25   0.5   0.01   0.05   0.1   0.99   0.8   1.25

1.3. Do the following calculations with Excel.

a. $1.96 * 2.5$

b. $2.5/\sqrt{40}$

c. $1.96*2.5/\sqrt{40}$

1.4. The following table, based on the American chamber of Commerce Researchers Association Survey for the second quarter of 2002, gives the prices of five items (in dollars) in 25 urban areas across the United States. Enter this table into Excel, format it similarly, and calculate the average of each column as follows: Insert Excel's AVERAGE function into the cell below the first column (or just type "=AVERAGE(" without the quotes), highlight the cells above it containing the values (creating cell references to the numbers), and then copy it to the cells below the other columns. *Save this file for future reference.*

City	Apartment Rent	Phone Bill	Price of Gasoline	Visit to Doctor	Price of Beer
Montgomery (AL)	$ 576	$ 22.28	$ 1.335	$ 52.33	$ 7.88
Juneau (AK)	1,020	18.26	1.584	88.67	8.12
Tucson (AZ)	689	21.03	1.347	54.80	7.79
Sacramento (CA)	749	16.99	1.643	70.00	6.99
San Diego (CA)	1,306	24.57	1.632	75.20	7.99
Denver (CO)	891	23.10	1.343	71.80	6.90
Hartford (CT)	896	22.39	1.419	80.25	7.15
Jacksonville (FL)	810	20.31	1.419	63.80	7.15
Bloomington (IN)	678	19.95	1.402	56.67	6.99
New Orleans (LA)	798	26.06	1.351	56.20	6.56
Boston (MA)	1,248	24.41	1.405	78.00	7.21
Grand Rapids (MI)	678	22.40	1.499	59.20	8.03
Minneapolis (MN)	815	25.16	1.366	72.20	7.49
Springfield (MO)	568	18.25	1.309	63.72	7.89
Billings (MT)	550	30.45	1.449	70.75	7.09
Buffalo (NY)	714	33.71	1.413	53.00	7.03
Charlotte (NC)	540	21.07	1.359	58.00	7.03
Akron (OH)	686	21.16	1.519	59.40	7.29
Oklahoma City (OK)	579	23.04	1.308	60.02	6.94
Portland (OR)	753	20.92	1.403	72.40	7.69
Philadelphia (PA)	1,282	21.12	1.360	62.50	8.57
Austin (TX)	1,025	19.20	1.299	68.33	6.78
Richmond (VA)	769	26.15	1.317	59.80	6.37
Spokane (WA)	593	18.49	1.305	61.80	6.89
Charleston (WV)	606	27.08	1.423	64.67	7.01

Explanation of variables:

Apartment Rent   Monthly rent of an unfurnished 2 bedroom apartment (excluding all utilities except water), 1 ½ or 2 baths, approximately 950 square feet.

Phone Bill   Monthly telephone charges for a private residential line (customer owns instruments).

Price of Gasoline   Price of one gallon regular unleaded, national brand.

Visit to Doctor   General practitioner's routine examination of patient.

Price of Beer   Heineken's 6-pack, 12-oz. Containers, excluding deposit.

CHAPTER **2**

Organizing Data

CHAPTER OUTLINE

2.1 SORTING RAW DATA

When data is collected and recorded, it is called "raw data." One of the first things you might want to do with a set of data is to sort it in numerical or alphabetical order. If you want to enter it into an Excel spreadsheet and have the computer sort it, then you need to enter the data *all into one column*. Excel can't sort it (all as a single variable) if the data is in rows *and* columns. There are two ways to sort the data. First, highlight the column.

Then, either go to **Edit>Sort** or click on one of the sorting icons: $\begin{smallmatrix}A\\Z\end{smallmatrix}\downarrow$ or $\begin{smallmatrix}Z\\A\end{smallmatrix}\downarrow$ (for reverse alphabetical or numerical order).

Example 2-1 A sample of 30 employees from large companies was selected, and these employees were asked how stressful their jobs were. The responses of these employees are recorded below where *very* represents very stressful, *somewhat* means somewhat stressful, and *none* stands for not stressful at all. Sort this data.

somewhat	none	somewhat	very	very	none
very	somewhat	somewhat	very	somewhat	somewhat
very	somewhat	none	very	none	somewhat
somewhat	very	somewhat	somewhat	very	none
somewhat	very	very	somewhat	none	somewhat

Solution:

	A	B	C	D	E	F			A	B	C	D
1	somewhat							1	none			
2	very							2	none			
3	very							3	none			
4	somewhat							4	none			
5	somewhat							5	none			
6	none							6	none			
7	somewhat							7	somewhat			
8	somewhat							8	somewhat			
9	very							9	somewhat			
10	very							10	somewhat			
11	somewhat							11	somewhat			
12	somewhat							12	somewhat			
13	none							13	somewhat			
14	somewhat							14	somewhat			
15	very							15	somewhat			
16	very							16	somewhat			
17	very							17	somewhat			
18	very							18	somewhat			
19	somewhat							19	somewhat			
20	somewhat							20	somewhat			
21	very							21	very			
22	somewhat							22	very			
23	none							23	very			
24	very							24	very			
25	none							25	very			
26	none							26	very			
27	somewhat							27	very			
28	somewhat							28	very			
29	none							29	very			
30	somewhat							30	very			

Figure 2.1 Raw data is entered into an Excel spreadsheet and then sorted (alphabetically).

2.2 ORGANIZING AND GRAPHING QUALITATIVE DATA

Once the data is sorted, it's easy to construct a frequency distribution table (where the data is grouped into its categories). Simply enter the categories in one column, and then count how many are in each category (this is the "frequency") and enter these counts into the second column. You can use Excel's COUNTIF function to count them for you, if you wish. Type the names of the categories into one column first. Click on the cell just to the right of the first category name. Then go to **Insert>Function** (or click on the f_x icon on the toolbar) and select COUNTIF. Highlight the column of data for the **Range** reference. Then click on the text box for the **Criteria** reference, and click on the cell just to the left of it that has the desired category name in it. Then hit **Enter** and voilá!

Example 2-2 Construct a frequency distribution for the data in the previous example.

Solution: Use Excel's COUNTIF function as described above and as shown below.

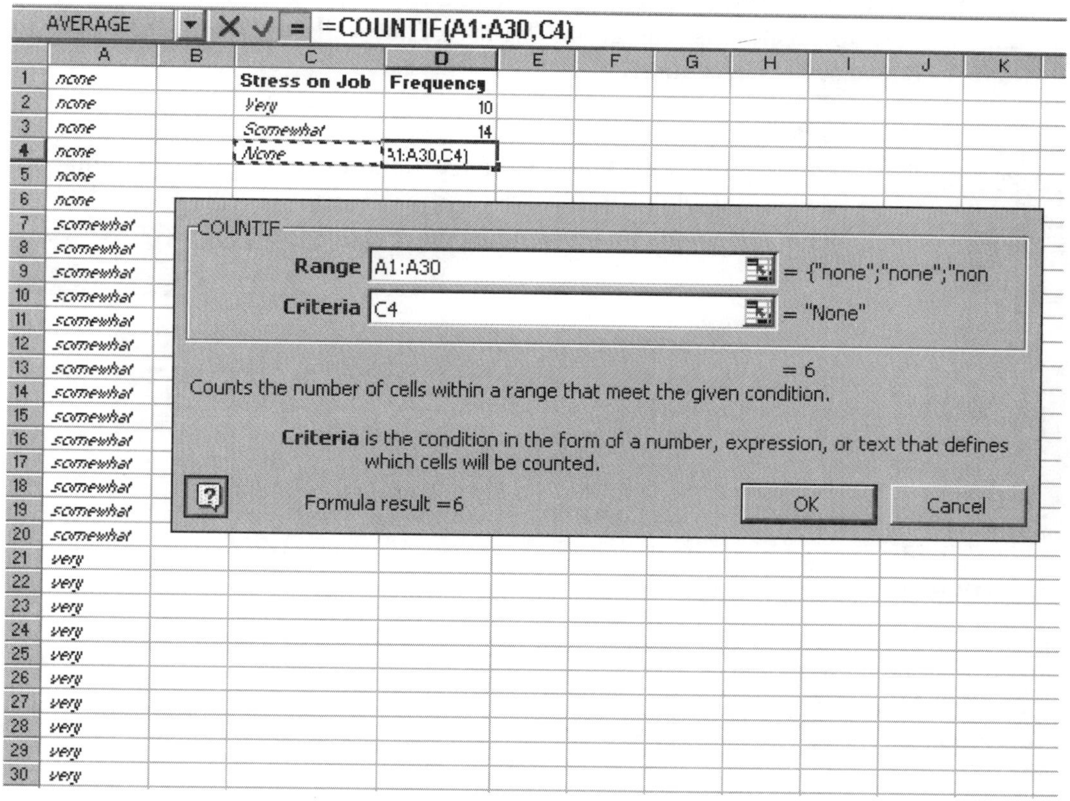

Figure 2.2 Using Excel's COUNTIF function to find frequencies of each category of data.

Example 2-3 Determine the relative frequency and percentage distributions for the data above.

Solution: In the cell next to the first frequency, type "=" and then click on the cell with the frequency. Then type "/" and enter the total of the frequencies. Then hit Enter. (The total of the frequencies can be found with the SUM function previously, if you wish.) This gives you the corresponding relative frequency. Since the same calculation is used for each relative frequency, it's easiest simply to copy it and paste it into the cells below. Notice how the cell references automatically change!

For the percentages, first copy the relative frequency values. You have to be careful here, though, since you don't want the cell references to change. Instead of doing a regular paste (like CTRL-V), you need to go to **Edit>Paste Special** and click on the radio button next to **Values.** (Another option is to make the cell references "absolute" by putting in $ signs before the column-letters and row-numbers. Here, it's quicker just to do a special paste.)

The percentages you want are really just the relative frequencies written as percents. So all you need to do next is change the format of the cells to percentages. Go to **Format>Cell** and select **Percentage.**

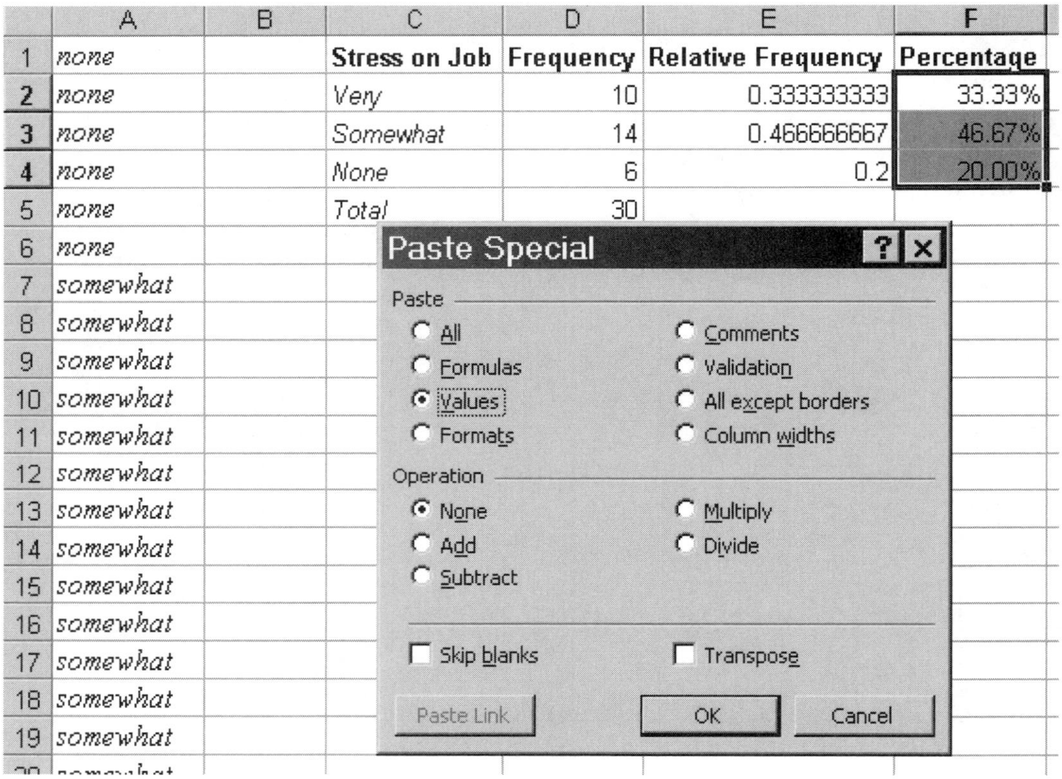

	A	B	C	D	E	F
1	*none*		**Stress on Job**	**Frequency**	**Relative Frequency**	**Percentage**
2	*none*		*Very*	10	0.333333333	33.33%
3	*none*		*Somewhat*	14	0.466666667	46.67%
4	*none*		*None*	6	0.2	20.00%
5	*none*		*Total*	30		
6	*none*					
7	*somewhat*					
8	*somewhat*					
9	*somewhat*					
10	*somewhat*					
11	*somewhat*					
12	*somewhat*					
13	*somewhat*					
14	*somewhat*					
15	*somewhat*					
16	*somewhat*					
17	*somewhat*					
18	*somewhat*					
19	*somewhat*					

Paste Special ? ✕

Paste
- ○ All
- ○ Formulas
- ● Values
- ○ Formats
- ○ Comments
- ○ Validation
- ○ All except borders
- ○ Column widths

Operation
- ● None
- ○ Add
- ○ Subtract
- ○ Multiply
- ○ Divide

☐ Skip blanks   ☐ Transpose

Paste Link   OK   Cancel

Figure 2.3 Using Excel's special pasting feature. (Cells have also been formatted to percentages.)

Bar Graphs and Pie Charts

Once you have a frequency distribution set up in Excel, it's really easy to make any kind of chart or graph that you wish from it. First, highlight the table of categories and frequencies (including the column headers). Then click on the "Chart Wizard" icon:

. The first step allows you to choose the type of chart that you would like. You can select either a vertical or horizontal bar chart, or you can select a pie chart, for example. Then there are several sub-types that you can choose from. Make your selections, and click on **Next.**

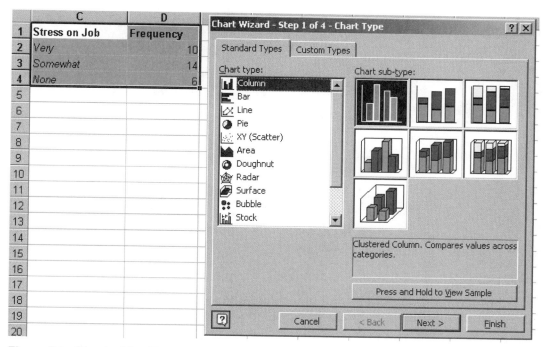

Figure 2.4 Step 1 of the Chart wizard. Select the Chart type and sub-type.

In step 2, you should see a preview of your chart. Here you need to make sure that it is interpreting your data correctly. For a vertical bar chart, the categories should be along the horizontal axis, and the frequencies along the vertical axis. If this is not the case, then you need to click on the **Series** tab and make corrections to what is being used for what. Then click on **Next** again.

In step 3, you get the opportunity to choose your chart title and axis labels. These will most likely need editing. The default chart title is often what you really want to put on your y-axis! See how they have been changed in the figure below. The other thing that you usually want to change is the legend. When one variable is being graphed, as in this example, you really don't need the legend. In fact, it detracts from the rest of the chart. To remove it, simply click on the **Legend** tab and uncheck the **Show legend** checkbox.

You can also edit things like the scales and fonts and orientations of the axes, and where and if gridlines are placed. But here, the default options are usually acceptable. Click on **Finish**, and you'll see your chart placed on the sheet. At this point you can move it (by clicking on it and dragging it to a new location) or resize it (by clicking on a corner, when the cursor turns to a diagonal arrow, and dragging it to a new size).

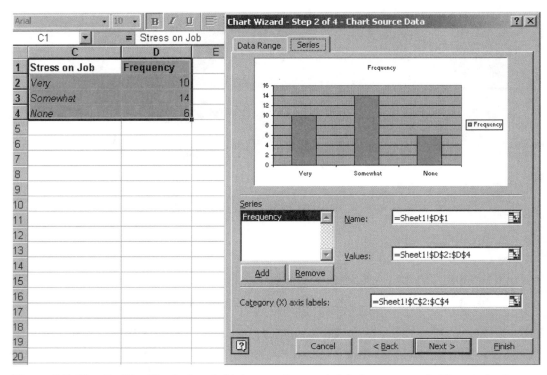

Figure 2.5 Step 2 of the Chart wizard. Make sure the x-axis labels are correct. The y-axis shows the "Values," which in this case are Frequencies.

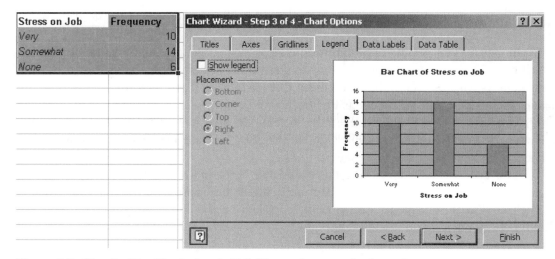

Figure 2.6 Step 3 of the Chart wizard. Edit titles and remove the legend.

If you want to change the width of the gap between the bars, you need to double-click on the bars themselves. Click on the **Options** tab and change the **Gap width.** If you want no gaps between the bars at all, type 0. (It can be any number between 0 and 500. The default is 150.)

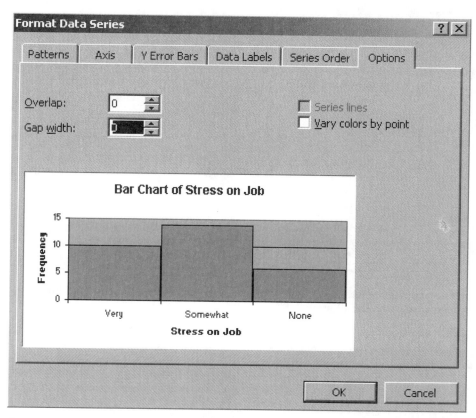

Figure 2.7   To change the width of the gap between bars, double-click on the bars themselves and then go to the **Options** tab. Here's where you can edit this feature.

A pie chart is just as easy. Select one of the pie chart options in step 1. Make sure the chart looks right in step 2. (Otherwise, you may need to change the data references.) If you have highlighted your frequency distribution before you click on the Chart Wizard, you should be able simply to click on **Next** at this point. In step 3, you should either a) leave the legend or b) delete it and select **Show label** under the **Data Labels** tab. It's also a good idea to have it show the percents. These are the relative frequencies which give the relative sizes of the pie pieces.

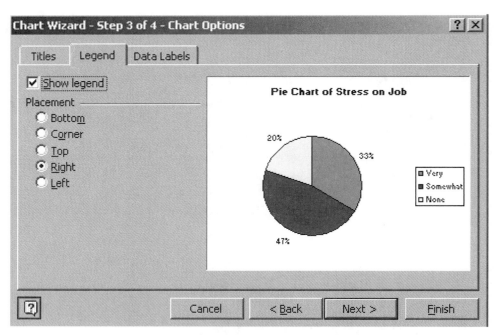

Figure 2.8 One good option for formatting a pie chart: keep the legend and click on the **Show percent** radio button under the **Data Labels** tab.

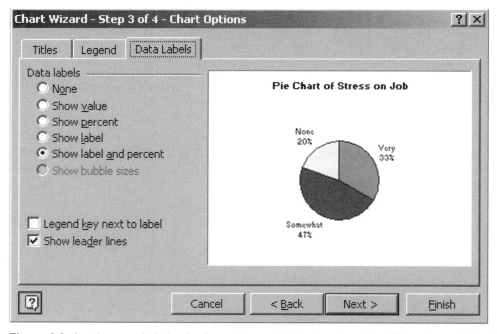

Figure 2.9 Another good choice for formatting a pie chart: uncheck the **Show legend** checkbox and have it show the labels with the percentages.

The last step, step 4, for creating any chart in Excel is to select the location for the chart. You can either place it as a resizable and movable object in the current worksheet, or you can create it as one large separate sheet. If you want to print it with the frequency distribution (and possibly other things on the worksheet) and have it take up less than a whole page, you should choose the first option. If you want to the chart to print large, as one whole page, then choose the second option.

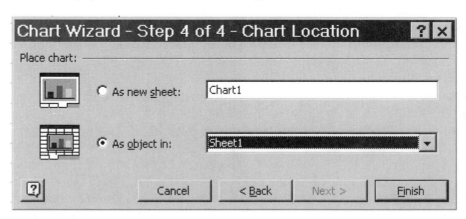

Figure 2.10 Select a location for your chart in the last step of the Chart Wizard.

2.3 ORGANIZING AND GRAPHING QUANTITATIVE DATA

In order to construct a frequency distribution for quantitative data, you first need to set up the classes (or subintervals) that you want to use. Then you can use Excel's FREQUENCY function to get the frequencies.

Example 2-4 The following data give the average travel time from home to work (in minutes) for 50 states. (Rows are alphabetical, by state.) The data are based on a sample survey of 700,000 households conducted by the Census Bureau (*USA TODAY*, August 6, 2001). Construct a frequency distribution for this data.

22.4	18.2	23.7	19.8	26.7	23.4	23.5	22.5	24.3	26.7	24.2
19.7	27.0	21.7	17.6	17.7	22.5	23.7	21.2	29.2	26.1	22.7
21.6	21.9	23.2	16.0	16.1	22.3	24.4	28.7	19.9	31.2	22.6
15.4	22.1	19.6	21.4	23.8	21.9	21.9	15.6	22.7	23.6	20.8
21.1	25.4	24.9	25.5	20.1	17.1					

Solution: Type the data *all into one column*, say column A, in an Excel worksheet. Highlight it and click on the sorting icon: $\overset{A}{\underset{Z}{\downarrow}}$, in order to sort it in numerical order. Notice the first value, which is now the minimum value of 15.4, and the last value, which is now the maximum value of 31.2. These will be used to divide the data up into classes. Suppose you want to group the data into six classes of equal width. Then you need to divide the range of the data (31.2-15.4) by 6. Click on cell C1 and type "=(31.2-15.4)/6" (without the quotes). You should see 2.633333 appear in the cell. Round this up to get your class width. You could use 2.7, but it would be nicer to use a class width of 3.0. Start the first class at some convenient value either equal to or just below the minimum. In this case, 15.0 is a nice round number just below the minimum value of 15.4.

Now what you need to do is set up two columns, one for the lower class limits and one for the upper class limits. They need to be separate in order to use Excel's FREQUENCY function. After you have your frequencies, you can edit the cells to put each entire class into one cell.

To start entering the lower class limits into column E, type 15 into cell E1. In the cell below it, cell E2, type "=" and then arrow up to cell E1 (for a cell reference to the value of 15), type "+3" (since 3 is the class width here) and then hit ENTER. You should see 18 appear in this cell. This will be the lower limit of the second class. Now just copy this cell down to the four cells (E3-E6) below it. These are all of your lower class limits.

Now enter the upper class limits into column F. The first *upper* class limit needs to be just below the second *lower* class limit. Since these data are accurate to one decimal place (the tenths place), you want the first upper class limit to be 0.1 (one tenth) less than the second lower class limit. So in cell F1, type "=18-0.1" (or you can click on the cell with the 18 in it instead of typing "18" if you wish). You should see 17.9 appear. Click in cell F2 right below it, type "=" and arrow up to cell F1, type "+3" and hit ENTER. Then copy this down to cells F3-F6. Each upper class limit should be exactly 0.1 less than the next class' lower class limit. (Data involving only whole numbers is easier; each upper class limit in that case is exactly 1 whole number less than the next class' lower class limit.)

Now you're ready to calculate the frequencies. Highlight cells G1-G6. Then go to **Insert>Function** (or click on the f_x icon on the toolbar) and select FREQUENCY. This is an "array" formula, so it's very important that you have *all six cells highlighted*. In fact, it wouldn't hurt to highlight cells G1-*G7*, since the formula will also count how many are above the last class (if there are any)! When the cursor is in the **Data_array** text box, highlight the column of original data. (You may need to move the FREQUENCY function's window out of the way in order to see your data. Simply click on it and drag it to where you want it.) Click in the **Bins_array** text box, and then highlight the *upper class limits.* (Whenever you see the term "bins" in Excel, think "upper class limits." These are the values that each class goes up to *and includes.*) Then, and again this is important and due to this being an "array" formula, instead of just hitting ENTER or clicking on **OK**, you need to hit CTRL-SHIFT-ENTER. Then, and only then, will you see all of the frequencies appear.

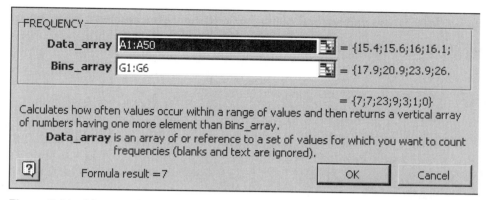

Figure 2.11   After entering the cells containing the data and the cells containing the upper class limits, hit CTRL-SHIFT-ENTER in order to see the frequencies show up in the highlighted column.

	G1	▼		=	{=FREQUENCY(A1:A50,F1:F6)}		
	A	B	C	D	E	F	G
1	15.4		2.633333		15	17.9	7
2	15.6				18	20.9	7
3	16				21	23.9	23
4	16.1				24	26.9	9
5	17.1				27	29.9	3
6	17.6				30	32.9	1
7	17.7						0
8	18.2						

Figure 2.12   The preliminary frequency distribution.

To finish the frequency distribution table, you need to combine the lower class limits with the upper class limits, putting a dash or the word "to" between them. Unfortunately, there is no nice easy way to do this. Probably the best approach is to copy the cells with the upper class limits and frequencies and then do a special paste, values only, into another location in the worksheet (or into a new sheet) so that you can then edit them. Highlight the cells with the upper class limits and frequencies in them (cells F1-G6) and hit CTRL-C or go to **Edit>Copy.** Click on a cell where you want the new table to appear or just click on the tab for Sheet2. Go to **Edit>Paste Special**, select **Values** and then click on **OK.** Now, simply edit the column of upper class limits to show the lower class limit to upper class limit intervals.

In order to edit a cell without losing its contents, you need to double-click on it. Then you'll be able to insert what you want. In this case, insert "15-" in front of "17.9" and "18-" in front of "20.9" and so on.

Finally, enter column headings. You may need to insert a row above your frequency distribution table in order to do this. Just click on the cell with the first class in it and go to **Insert>Rows**. Then in the cell above the classes, type a label for the classes, like "Average Travel Time to Work (minutes)" and in the cell above the frequencies, type

"Frequency." So that you can see everything nicely, highlight the entire table, go to **Format>Column>AutoFit Selection.** Then center it by clicking on the center icon: ≣.

	A	B
1	Average Travel Time to Work (minutes)	Frequency
2	15-17.9	7
3	18-20.9	7
4	21-23.9	23
5	24-26.9	9
6	27-29.9	3
7	30-32.9	1

Figure 2.13 The finished frequency distribution.

Calculating relative frequencies can be done the same way as in the previous section, by dividing the frequencies by their total. To get their total, click on the cell below them, type "=SUM(" (or click on the f_x icon on the toolbar and select SUM), highlight the cells with the frequencies in them, and hit ENTER (or click on **OK**). Then in the cell to the right of the first frequency, type "=" and arrow left (to reference that frequency), type "/" and then the sum value and hit ENTER. Copy this down to the cells below it and you should see all of the corresponding relative frequencies. Format them to percentages if you wish.

Histograms and Frequency Polygons

As discussed in the previous section on qualitative data, once you have your frequency distribution set up in Excel, it's really easy to make a chart or graph from it. A histogram is really just a vertical bar chart for quantitative data. Just as before, you first need to highlight the table of classes and frequencies (including the column headers). Then click on the "Chart Wizard" icon: ▥. Select the vertical bar chart. Follow the same 4 steps as noted in the previous section, entering an appropriate title and axis labels and removing the legend. Make sure that you also remove the gaps between the bars. A histogram shows classes that have no gaps between them, so therefore the bars, which represent these classes, should have no gaps between them. Recall that to do this you must double-click on the bars themselves (on the finished chart), select the **Options** tab, and then enter a **Gap width** of zero. (Any gap between bars on a histogram should really represent a class with zero frequency.)

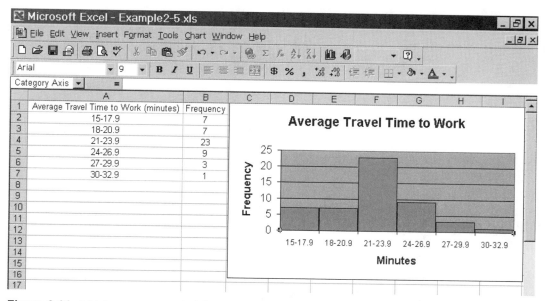

Figure 2.14 A histogram generated from the frequency distribution by using the Column chart option in Excel.

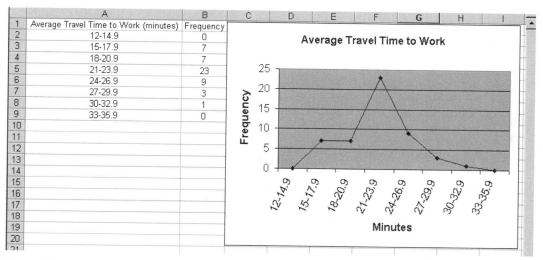

Figure 2.15 A frequency polygon generated from the frequency distribution by using the Line chart option in Excel.

In order to have Excel generate a frequency polygon instead of a histogram, simply select the **Line** chart option in step 1 instead of the **Column** chart option. Otherwise, the steps are the same. However, you will need to add two empty classes to your frequency distribution *first*, one just above your first class (insert a row in order to do this) and one just below your last class. Remember that consecutive lower class limits and consecutive upper class limits need to be exactly a class width (3, in this case) apart. So the classes added to the above table are 12-14.9 and 33-35.9. Enter frequencies of 0 for both of these classes. Then highlight this table and go to the Chart Wizard.

2.4 USING EXCEL'S DATA ANALYSIS TOOLPAK

Another option for generating frequency distributions and histograms is to use Excel's Data Analysis ToolPak. Go to the **Tools** menu and see if **Data Analysis** is an option. If not, then go to **Tools>Add-Ins.** Check the checkboxes next to **Analysis ToolPak** and **Analysis ToolPak -VBA** and click on **OK.** From now on, it will be loaded when Excel is opened. Go to **Tools** and you should now see **Data Analysis.** (If this tool was not originally installed with Excel, then you will need to reinstall Excel in order to have it available.)

In order to use the Data Analysis tool to create your frequency distribution and histogram, you still need to have your upper class limits (what Excel calls "bins") set up as before. Then go to **Tools>Data Analysis** and select the **Histogram** option. For the **Input Range**, highlight the original data. For the **Bins Range**, highlight the upper class limits. Then check the checkbox next to **Chart Output** and click on **OK.**

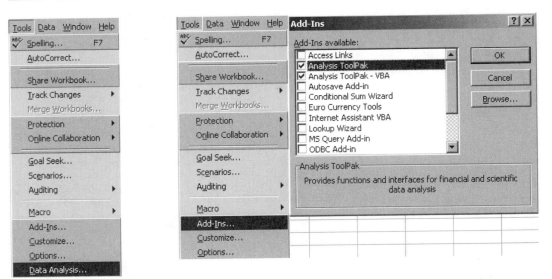

Figure 2.16 Go to **Tools>Data Analysis** in order to access the ToolPak. If it is not listed, go to **Tools>Add-Ins** and select the two checkboxes so that it will always be an option in the future.

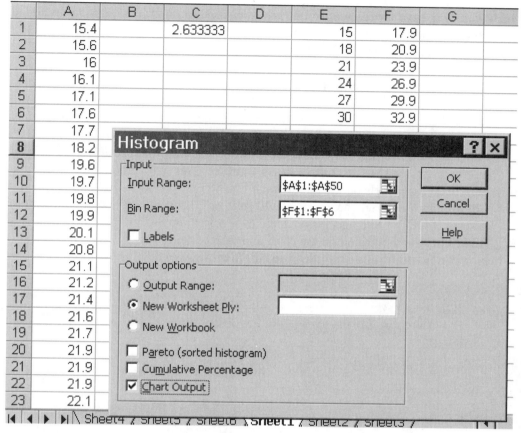

Figure 2.17 Using Excel's Data Analysis ToolPak to generate a frequency distribution and histogram.

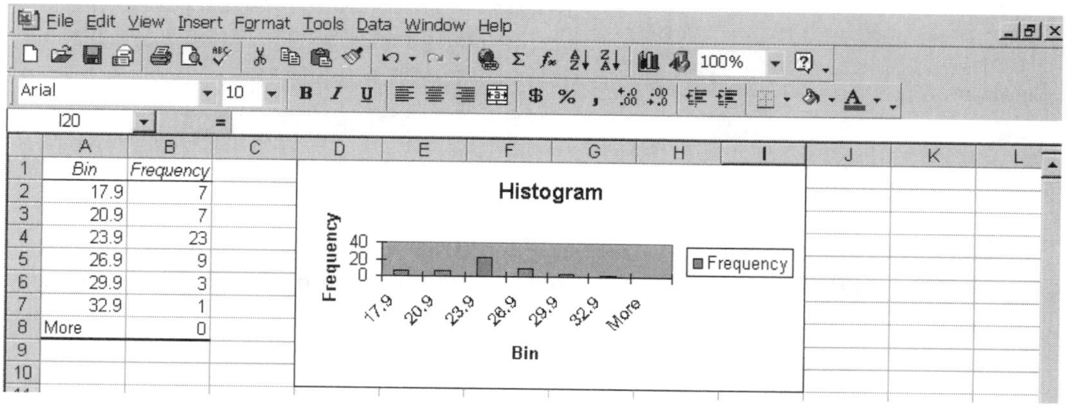

Figure 2.18 The preliminary frequency distribution and histogram generated by Excel's Data Analysis ToolPak that need a good bit of editing.

Now you will have a *preliminary* frequency distribution and histogram in an added worksheet. These need lots of work, though! Here's what you need to do to fix them:

1. Resize the histogram by clicking on it and dragging a corner so that you can see the bars well.
2. Double-click on the bars, choose the **Options** tab, and change the gap width to zero.
3. Right-click on the row number that contains the "More" class in the frequency distribution (which should have a frequency of zero) and select **Delete.**
4. Edit the "bins" so that they show the entire lower class limit to upper class limit intervals (instead of just upper class limits). This will automatically fix the horizontal axis labels on the histogram.
5. Change the column heading of "Bins" to "Classes" or something more descriptive, like "Average Travel Time to Work (minutes)" in this case.
6. Right-click on the legend on the chart and select **Clear.**

Click on the chart title and axis titles and edit them.

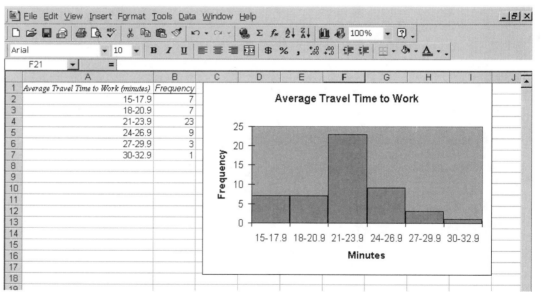

Figure 2.19 The "fixed" frequency distribution and histogram.

Excel has several "automatic" features like this, which can save time but often need some clean-up work. You can decide which you prefer: creating things from scratch or fixing what Excel has created for you!

2.5 CUMULATIVE FREQUENCY DISTRIBUTIONS

In order to change a frequency distribution to a *cumulative* frequency distribution, you need to change all of the lower class limits to the same first lower class limit, and then calculate the cumulative frequencies.

Example 2-5 Construct a cumulative frequency distribution for the 50 states' average travel times given in Example 2-4.

Solution: First, construct the frequency table as above, but instead of editing the classes to show varying lower class limits with the upper class limits, use 15.0 as the lower class limit for *all* of the classes. (These will be *cumulative* classes.)

In the column next to the frequencies, type in the heading "Cumulative Frequency." For the first one, just copy the first class' frequency. For the second one, type "=" and then arrow left, and type "+" and then arrow up to the previous cumulative frequency. Then copy this cell down to the cells below it (click on it and drag the square in the bottom right-hand corner).

Finally, hide the column with the frequencies in it. (You can't delete it, because then the references are gone, but you can hide it.) Right-click on the column letter and select **Hide**.

	C7	▼	=	=B7+C6
	A		**C**	
1	Cumulative classes		Cumulative Frequency	
2	15-17.9		7	
3	15-20.9		14	
4	15-23.9		37	
5	15-26.9		46	
6	15-29.9		49	
7	15-32.9		50	

Figure 2.20 A cumulative frequency distribution. The cumulative frequencies are calculated by using cell references to frequencies in a hidden column.

In order to generate an Ogive, which is really a cumulative frequency polygon, you need to enter upper class *boundaries* into a column. The upper class boundary of a class is always halfway between the upper class limit of that class and the next class' lower class limit. So you can have Excel average these values for you, if you wish. Then copy the cumulative frequencies into the column next to them. (Go to **Edit>Paste Special** and select **Values** in order to avoid changing cell references.)

Finally, insert a row above the first upper class boundary, enter the number that is exactly one class width less than the first upper class boundary (in the cell just above it), and enter a 0 in the cell just above the first cumulative frequency.

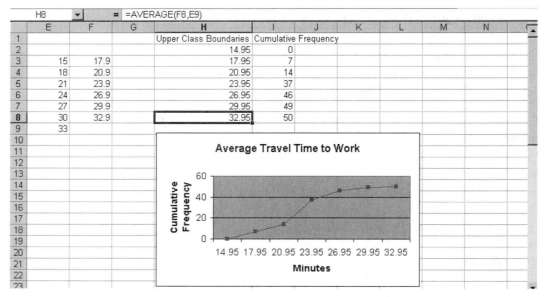

Figure 2.21 Calculate upper class boundaries and position cumulative frequencies next to them in order to generate an Ogive.

Highlight the columns and go to the Chart Wizard. Select the **Line** chart option. In step 2, you will probably need to fix the data references. Click on the **Series** tab. Under **Series**, you need only to have the Cumulative Frequency series. Remove the other one. Click in the text box next to **Category (X) axis labels** and highlight the column of upper class boundaries. Then you should see a good preview of the ogive. Remove the legend and edit titles in step 3, and click on **Finish.**

2.6 STEM-AND-LEAF DISPLAYS

A stem-and-leaf display can be done in Excel just slightly more easily than it can be done by hand since Excel can sort the data for you. Enter the stems into one column, and make sure they are right-justified. Then enter the leaves in the next column, next to the stems, and make sure they are left-justified. Finally, format the vertical border between the two cells to be a dark thick line.

Example 2-6   The following are the scores of 30 college students on a statistics test. Construct a stem-and-leaf display.

75	52	80	96	65	79	71	87	93	95
69	72	81	61	76	86	79	68	50	92
83	84	77	64	71	87	72	92	57	98

Solution: Enter the data all into one column, say column A, in an Excel Spreadsheet and sort it. Notice the tens digits that are involved; these are your stems. Type these digits into another column, say column C. Since these are numbers, Excel should automatically keep them in the far right of each cell. If not, highlight this column and click on the right-justify icon: ☰. In the cell next to each stem, then, type the corresponding ones digits, or leaves, with a space between them. Since the spaces cause them to be formatted as text, Excel should automatically keep them to the far left of the cell. But again, if not, you can highlight this column and click on the left-justify icon: ☰. Finally, with this column highlighted, right-click and choose **Format Cells.** Click on the **Border** tab. Select a dark line style on the right side of this window and then click on the left border of the picture shown on the left. Then click **OK.** You should now have a nice stem-and-leaf display of this data!

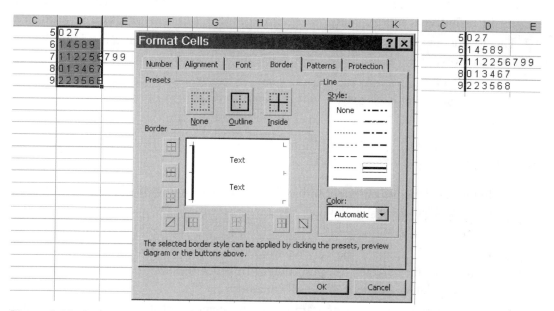

Figure 2.22 Creating a stem-and-leaf display in Excel. The last step is to format the cells to have a bold dividing line.

One of the advantages of a stem-and-leaf plot is its simplicity, though. You really don't need Excel to do this for you!

Exercises

2.1 Fifty students selected from a university were asked about their student status, and the results are shown below. F, SO, J, and SE are the abbreviations for freshman, sophomore, junior, and senior, respectively.

J	F	SO	SE	J	J	SE	J	J	J
F	F	J	F	F	F	SE	SO	SE	J
J	F	SE	SO	SO	F	J	F	SE	SE
SO	SE	J	SO	SO	J	J	SO	F	SO
SE	SE	F	SE	J	SO	F	J	SO	SO

a. Enter this data into an Excel spreadsheet (all in one column) and sort it.

b. Prepare a frequency distribution table using Excel.

c. Have Excel calculate the relative frequencies for all categories. Copy these values and format the copies as percentages in order to show percentages for all categories as well.

d. Have Excel generate a bar graph for the frequency distribution. Label titles and axes appropriately.

e. Have Excel generate a pie chart for the (relative) frequency distribution. Label titles and axes appropriately.

2.2 The 50 students selected from a university were also asked their ages, and this data is shown below.

21	19	24	25	29	34	26	27	37	33
18	20	19	22	19	19	25	22	25	23
25	19	31	19	23	18	23	19	23	26
22	28	21	20	22	22	21	20	19	21
25	23	18	37	27	23	21	25	21	24

a. Enter this data into an Excel spreadsheet (all in one column) and sort it.

b. Prepare a frequency distribution table using Excel that has 7 classes of equal width.

c. Have Excel calculate the relative frequencies for all categories. Copy these values and format them as percentages in order to show percentage as well as decimal relative frequencies for all categories.

d. Have Excel generate a histogram for the frequency distribution. Label titles and axes appropriately. Remove the legend and remove the gaps between the bars.

e. Have Excel generate a frequency polygon. Label titles and axes appropriately.

f. Prepare a cumulative frequency distribution.

g. Have Excel generate an ogive for the cumulative frequency distribution.

2.3. Follow steps b-g above for each of the columns of data given in Exercise 1.4 (the prices of the five items in 25 urban areas across the United States). Use one sheet for each item. Rename the worksheets with the names of the five items.

CHAPTER **3**

Numerical Descriptive Measures

CHAPTER OUTLINE

3.1 Measures of Central Tendency
3.2 Measures of Dispersion
3.3 Finding Measures for Grouped Data

3.4 Using Excel's Data Analysis ToolPak
3.5 Measures of Position

3.1 MEASURES OF CENTRAL TENDENCY

Excel has many of the numerical descriptive measures used in statistics, in particular the measures of central tendency, as built-in functions. Recall that to access any of these built-in functions, you need to go to **Insert>Function** or click on the f_x icon on the toolbar. Either click on **All** in order to scroll through all of Excel's functions, (they're listed in alphabetical order), or select the **Statistical** category to restrict the list. In order to find the mean of a set of data, use the AVERAGE function. For the median, use the MEDIAN function.

Example 3-1 The following are the ages of all eight employees of a small company.

| 53 | 32 | 61 | 27 | 39 | 44 | 49 | 57 |

Find the mean and median age.

Solution: Type the ages into an Excel spreadsheet. (If you want to sort them, make sure you type them all into one column. Otherwise, it doesn't matter how you enter the data.) Click on an empty cell near the data. Click on the f_x icon, select the AVERAGE function, and click **OK**. Move the window out of the way, if necessary, so that you can then highlight the ages (for the **Number1** text box). Then click on **OK**. You should see the mean age of 45.25 appear.

A3	▼	=	=AVERAGE(A1:H1)					
	A	B	C	D	E	F	G	H	I
1	53	32	61	27	39	44	49	57	
2									
3	45.25								
4									

Figure 3.1 Using Excel's AVERAGE function to find the mean of a set of data.

Another option is simply to type "=AVERAGE(" instead of clicking on the f_x icon and selecting this function. Then highlight the data and hit ENTER.

For the median, click on another cell near the data, click on the f_x icon, and select the MEDIAN function (or type "=MEDIAN(" without the quotes). Again highlight the data, moving the function window out of the way if necessary, and hit ENTER. You should see the median age of 46.5 appear.

C3	▼	=	=MEDIAN(A1:H1)					
	A	B	**C**	D	E	F	G	H	I
1	53	32	61	27	39	44	49	57	
2									
3	45.25		46.5						
4									

Figure 3.2 Using Excel's MEDIAN function to find the median of a set of data.

Excel also has a MODE function built in, but you have to be very careful with this function, because *it only finds the **first** mode* if there are several. Sorting the data in ascending order and then in descending order will alert you to the existence of another mode. Then you need to find any others manually. However, the MODE function does clearly indicate when there is no mode.

Example 3-2 For the ages of the eight employees from Example 3-1, find the mode(s).

Solution: After entering the data into Excel, click on a cell near it, type "=MODE(" and highlight the data, and hit ENTER.

E3	▼	=	=MODE(A1:H1)					
	A	B	C	D	**E**	F	G	H	I
1	53	32	61	27	39	44	49	57	
2									
3	45.25		46.5		#N/A				
4									

Figure 3.3 Excel's MODE function indicates that there is no mode for this set of data. (Each value appears only once.)

In this case, there is no mode. If there is a mode, you need to sort the data to make sure that you notice whether there is one mode or *more* than one mode.

Example 3-3 The following data give the speeds (in miles per hour) of eight cars that were stopped on I-95 for speeding violations.

| 77 | 69 | 74 | 81 | 71 | 68 | 74 | 73 |

Find the mode(s).

Solution: Type the data into an Excel spreadsheet, *all into one column* so that you can sort it. Click on any blank cell near the data, click on the f_x icon and select the MODE function (or type "=MODE(" without the quotes), highlight the data, and hit ENTER. You should see a mode of 74 appear. Now, to make sure that this is the only mode, highlight the data and sort it in ascending order by clicking on the ![A/Z down arrow] sorting icon, and then sort it in descending order by clicking on the ![Z/A down arrow] sorting icon. *Watch the mode value while you do this!* If it changes, then there is more than one mode and you need to inspect your data in order to find any others. If it stays the same, as in this case, then you know that there is only one mode.

Example 3-4 The prices of the same brand of television set at eight stores are found to be $495, $486, $503, $495, $470, $505, $470, and $499. Find the mode(s).

Solution: Follow the same directions as above. In this case, you should see the mode change from $470 to $495 when the data is sorted in ascending order and then descending order. After inspecting the data, you can see that these are the only two values that both appear twice; the others appear only once.

C1		= =MODE(A1:A8)				C1		= =MODE(A1:A8)		
	A	B	C	D	E		A	B	C	D	E
1	470		470			1	505		495		
2	470					2	503				
3	486					3	499				
4	495					4	495				
5	495					5	495				
6	499					6	486				
7	503					7	470				
8	505					8	470				

Figure 3.4 When there is more than one mode, the value calculated by the MODE function changes as the data is sorted in ascending and then descending order.

Example 3-5 The ages of 10 randomly selected students from a class are 21, 19, 27, 22, 29, 19, 25, 21, 22, and 30. Find the mode(s).

Solution: Again, follow the same steps as noted in the previous two examples. When the data is unsorted, you will see the mode of 21 appear. (That's because 21 is listed first, and Excel's MODE function always finds only the *first* mode listed.) When the data is sorted in ascending order, you'll see the mode of 19 appear. And when it is then sorted in descending order, you'll see the mode of 22! Look carefully at the data to see that these are all of the modes.

Excel's MODE function does *not* work for qualitative data given as text instead of numerical values, unfortunately. In this case, it is best to use Excel to sort the data in alphabetical order, and then manually inspect it.

3.2 MEASURES OF DISPERSION

In order to find the range of a set of data, you can use Excel's MIN and MAX functions along with subtraction.

Example 3-6 Following are the 2002 earnings (in thousands of dollars) before taxes for all six employees of a small company. Find the range.

| 48.50 | 38.40 | 65.50 | 22.60 | 79.80 | 54.60 |

Solution: Enter the data into Excel. Click on a blank cell near the data. Click on the f_x icon, select MAX, highlight the data (moving the window out of the way if necessary), and click on **OK.** You should see the maximum salary of 79.8 (for $79,000) appear. Then click in the formula bar right after the last (closed) parenthesis, type "-" and click on the f_x icon to select MIN. Highlight the data again and hit ENTER (or click on **OK**). You should see the range of 57.2 (for $57,200) appear. The other option is simply to type the whole formula into the cell. You can even type the cell references to the data, if you prefer, instead of highlighting the data.

C1		▼		=	=MAX(A1:A6)-MIN(A1:A6)		
	A	B	C	D	E	F	
1	48.5		57.2				
2	38.4						
3	65.5						
4	22.6						
5	79.8						
6	54.6						

Figure 3.5 Use Excel's MAX and MIN functions with subtraction in order to find the range.

In order to calculate the variance and standard deviation, you first need to observe whether the data represents a sample or a population. If it is sample data, use the functions VAR and STDEV. If it is census data (from the entire population), use the functions VARP and STDEVP.

Example 3-7 Find the variance and standard deviation for the data in the previous example.

Solution: Because the data are the earnings of *all* employees of this company, we use the population functions. Enter the data into Excel, click on a blank cell nearby, select VARP, highlight the data, and hit ENTER. You should see the variance of 337.0489 appear. Click on another blank cell nearby, select STDEVP, highlight the data, and hit ENTER. You should see the standard deviation of approximately 18.359 (for $18,359) appear. (Or, since the standard deviation is the square root of the variance, you could just use the SQRT function and refer to the cell with the variance in it!)

C5		=	=STDEVP(A1:A6)		
	A	B	C	D	E
1	48.5	Range:	57.2		
2	38.4				
3	65.5	Variance:	337.0489		
4	22.6				
5	79.8	Standard Deviation:	18.35889		
6	54.6				

Figure 3.6 Use Excel's functions VARP and STDEVP to find the variance and standard deviation of data from a population.

Example 3-8 Find the variance and standard deviation of the ages of the 10 randomly selected students from a class given in Example 3-5.

Solution: This data is from a sample, since the students were randomly selected from a class of presumably more than 10 people. Follow the same directions as above, but this time select the functions VAR and STDEV (or type each in with an "=" before and a "(" after). You can also label each if you wish, as was done in the figure above and the figure below.

D3	▼	=	=STDEV(A1:A10)	
	A	B	C	D	E
1	21		Variance:	16.05556	
2	19				
3	27	Standard Deviation:		4.006938	
4	22				
5	29				
6	19				
7	25				
8	21				
9	22				
10	30				

Figure 3.7 Use Excel's VAR and STDEV functions to find the variance and standard deviation of data taken from a sample.

3.3 FINDING MEASURES FOR GROUPED DATA

In the previous two sections, we found the measures of central tendency and dispersion for "ungrouped" data, where it was simply given as a list. In this section, we'll look at how to use Excel to find the mean, variance, and standard deviation from "grouped" data, where values are given in groups, as in a frequency distribution table.

Unfortunately, Excel does not have an easy "one-step" function for this situation, but it definitely provides some advantages over doing it by hand or even with a calculator.

If you have a frequency distribution table with interval classes, you first need to calculate the class midpoints. You can use Excel's AVERAGE function for this, since class midpoints are simply the averages of the lower and upper class limits. After that, there are two possible approaches. One approach is to use the columns and copying feature in Excel in order to calculate the mean, variance, and standard deviation using the formulas in your textbook.

Example 3-9 The table below gives the frequency distribution of the number of orders received each day during the past 50 days at the office of a mail-order company. Use Excel and the formulas in your textbook to calculate the mean, variance, and standard deviation.

Number of Orders	Number of Days
10-12	4
13-15	12
16-18	20
19-21	14

Solution: In column A of an Excel worksheet, calculate the midpoints of the classes by typing "=(10+12)/2" and hitting ENTER for the first one, then "=(13+15)/2" for the second one, and so on. In column B, next to these midpoints, enter the frequencies.

In column C, have Excel calculate the products of the midpoints and frequencies. For the first one, click on cell C1, type "=" and click on the first midpoint (cell A1), then type "*" and click on the first frequency (cell B1), and then hit ENTER. You should see 44 appear. Now just copy this first one down to the cells below it. (Click on cell C1 and drag the square in the bottom right corner down to cell C4.) Remember that the cell references adjust themselves! These products will be used for the mean.

In column D, you need to have Excel calculate the products of the squares of the midpoints times the frequencies. These will be used for the variance and then standard deviation. To calculate the first one, click on cell D1, type "=" and click on the first midpoint (cell A1), then type "^2*" (to raise it to the second power and then multiply) and click on the first frequency (cell B1), and hit ENTER. You should see the value of 484 appear. Now copy this down to the cells below it (cells D2-D4).

Now we need some totals. In cell B6, below the frequencies (leaving a row space), type "=SUM(" (or click on the f_x icon and select the SUM function), highlight the cells containing the frequencies, and hit ENTER. You should see 50 appear. This is the size of the data set! Now copy this cell to cells C6 and D6 to get the sums of those columns as well. (You can drag it to the right, if you wish.)

To calculate the mean of this data, you need to have Excel divide the sum of column C (the value in cell C6) by the sum of the frequencies in column B (the value in cell B6).

	B8	▼	=	=C6/B6
	A	B	C	
1	11	4	44	
2	14	12	168	
3	17	20	340	
4	20	14	280	
5				
6		50	832	
7				
8	Mean:	16.64		

Figure 3.8 One method for calculating the mean of grouped data: midpoints are calculated in column A, frequencies are entered in column B, products of midpoints and frequencies are calculated in column C, and then the sum of column C is divided by the sum of column B.

To have Excel calculate the variance, you need to notice that the data is from a sample. Therefore, you need to take the sum of column D, subtract the square of the sum of column C divided by the sum of column B from it, and then divide that difference by *one less than* the sum of column B. This formula is shown in the formula bar in the figure below.

To calculate the standard deviation, then, all you need to do is take the square root of the variance. Type "=SQRT(" into an empty cell (or click on the f_x icon and select the SQRT function), click on the variance, and hit ENTER.

	E8	▼		=	=(D6-C6^2/B6)/(B6-1)			
	A	B	C	D	E	F	G	H
1	11	4	44	484				
2	14	12	168	2352				
3	17	20	340	5780				
4	20	14	280	5600				
5								
6		50	832	14216				
7								
8	Mean:	16.64		Variance:	7.582041	Standard Deviation:	2.753551	

Figure 3.9 Using Excel to calculate the variance via columns, sums, and a mathematical formula. The standard deviation is then calculated using Excel's built-in SQRT function.

Another, perhaps easier and quicker, approach to calculating the mean, variance, and standard deviation of grouped data is to *copy each midpoint as many times as its frequency indicates* and then just use the AVERAGE, VAR and STDEV (or VARP and STDEVP if the data is from a population) functions!

Example 3-10 Use copies of the midpoints of the classes in the frequency distribution table in Example 3-9 and Excel's built-in functions in order to calculate the mean, variance, and standard deviation.

Solution: Calculate the midpoints in a blank Excel worksheet as before. Notice that the frequency of the first class is 4. Type the first midpoint, 11, into cell C1 and copy it down to cells C2-C4. Notice that the frequency of the second class is 12. Type the second midpoint, 14, into cell D1 and copy it down to cells D2-D12. Type the third midpoint of 17 into cell E1 and copy it down to cells E2-E20, since its frequency is 20. Finally, type the fourth midpoint of 20 into cell F1 and copy it down to cells F2-F14, since its frequency is 14.

Now just insert the AVERAGE, VAR, and STDEV functions, one at a time, into nearby blank cells, say cells H1, H3, and H5, highlighting cells C1-F20 (so that all of the data is included) for each. The blank cells around the copied midpoints that are included in this range will be ignored. Excel automatically calculates all three of these measures for you! Label them if you wish.

For population data, proceed this same way, but use the AVERAGE, VARP, and STDEVP functions instead.

	A	B	C	D	E	F	G	H
1	11		11	14	17	20	mean:	16.64
2	14		11	14	17	20		
3	17		11	14	17	20	variance:	7.582041
4	20		11	14	17	20		
5				14	17	20	st. dev.:	2.753551
6				14	17	20		
7				14	17	20		
8				14	17	20		
9				14	17	20		
10				14	17	20		
11				14	17	20		
12				14	17	20		
13					17	20		
14					17	20		
15					17			
16					17			
17					17			
18					17			
19					17			
20					17			

Figure 3.10 Copy the midpoints each as many times as its frequency indicates and then you can use Excel's built-in functions, AVERAGE, VAR (or VARP for population data), and STDEV (or STDEVP for population data) in order to calculate these measures for grouped data.

3.4 USING EXCEL'S DATA ANALYSIS TOOLPAK

For ungrouped data *that is from a sample*, you can also use Excel's Data Analysis ToolPak to find all of the measures of central tendency and dispersion (and more) all at once.

Example 3-11 Use Excel's Data Analysis ToolPak to find the mean, the median, a mode, the range, the variance and the standard deviation of the ages of the 10 randomly selected students from a class given in Example 3-5.

Solution: Enter the data into an Excel worksheet. (See Figure 3.7.) Go to **Tools>Data Analysis.** (If this is not listed, then you need to go to **Tools>Add-Ins**, check the checkboxes next to **Analysis ToolPak** and **Analysis ToolPak –VBA**, and click on **OK.** From then on, it will be loaded when Excel is opened. Go to **Tools** and you should now see **Data Analysis.** If this tool was not originally installed with Excel, then you will need to reinstall Excel in order to have it available.)

When the first window pops up, select **Descriptive Statistics.** When the next window pops up, highlight the cells that contain the data for the **Input Range**. Then check the checkbox next to **Summary statistics.** When you click on **OK** or hit ENTER, a new worksheet is created with some text in column A and values in column B. So that you can read it, go to **Format>Column>AutoFit Selection** while the columns are still highlighted.

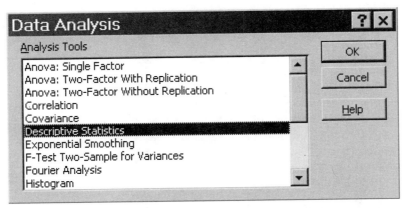

Figure 3.11 Descriptive Statistics is one of the options in Excel's Data Analysis ToolPak.

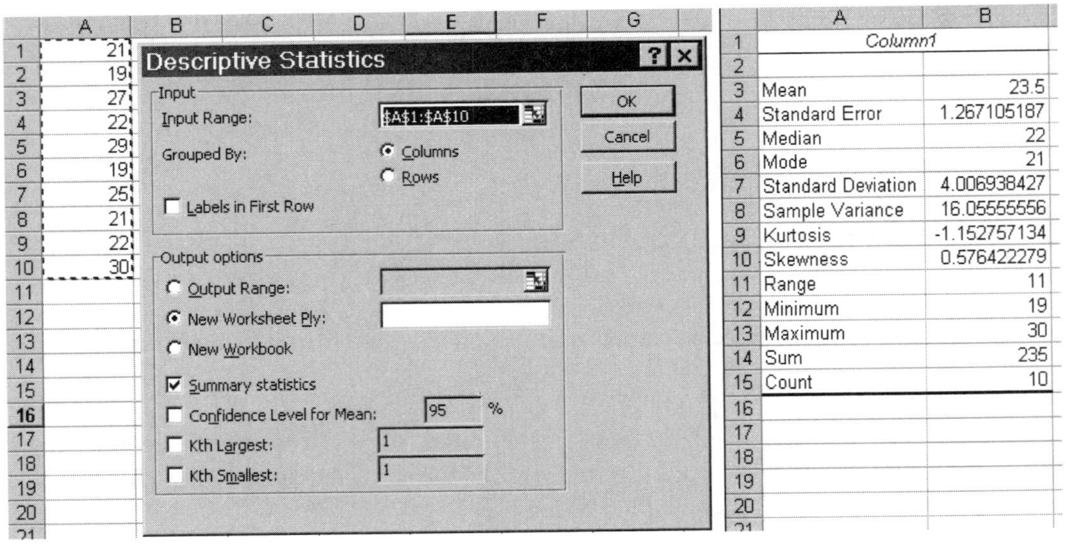

Figure 3.12 Highlight the data and check the **Summary statistics** checkbox in order to generate the summary statistics in a new worksheet. Increase column widths in order to read the text.

Notice that all in this one step, you get (among other things) the mean, the median, *one of the modes*, the standard deviation, the variance, and the range! Just be careful about the mode, since you have to go back to the data (and, if you need to, the MODE function together with sorting the data forwards and backwards) in order to see if there is more than one mode.

The main disadvantage of getting the measures this way is that you can't see the reference back to the original data. When you use individual functions, those cell references are always preserved. Here, they are not.

3.5 MEASURES OF POSITION

Quartiles and percentiles for a set of data can easily be found with Excel's QUARTILE and PERCENTILE functions. These functions require *two* inputs: 1) the array of data and 2) the particular quartile or percentile desired. If you insert one of these functions by going to **Insert>Function** or by clicking on the f_x icon, then you will see two text boxes for these two inputs, both of which must be filled in. For the **Array** text box in either function, simply highlight the data as you've done before. Then click in the second text box and enter the second input. If you want to type in the function name (with an "=" before it and a "(" after it), then after highlighting the data, you need to type a comma before entering the second input.

The second input for the QUARTILE function, **Quart**, must be a 0, 1, 2, 3 or 4. Entering a 0 will give you the minimum of the data, and entering a 4 will give you the maximum of the data. Enter a 1, 2, or 3 for the 1st, 2nd, or 3rd quartile, respectively.

The second input for the PERCENTILE function, **K,** must be a decimal between 0 and 1. Again, if you enter a 0, you'll get the minimum of the data set. If you enter a 1, you'll get the maximum value of the data set. Otherwise, you need to enter the decimal that corresponds to the percentage of the percentile desired. For example, the 90th percentile would be found by entering a value of 0.90.

Example 3-12 The table below lists the total revenues for the 12 top-grossing North American concert tours of all time. Use Excel to find the values of the three quartiles.

Tour	Artist	Total Revenue (millions of dollars)
Steel Wheels, 1989	The Rolling Stones	98.0
Magic Summer, 1990	New Kids on the Block	74.1
Voodoo Lounge, 1994	The Rolling Stones	121.2
The Division Bell, 1994	Pink Floyd	103.5
Hell Freezes Over, 1994	The Eagles	79.4
Bridges to Babylon, 1997	The Rolling Stones	89.3
Popmart, 1997	U2	79.9
Twenty-Four Seven, 2000	Tina Turner	80.2
No Strings Attached, 2000	'N-Sync	76.4
Elevation, 2001	U2	109.7
Popodyssey, 2001	'N-Sync	86.8
Black and Blue, 2001	The Backstreet Boys	82.1

Source: Fortune, September 30, 2002

Solution: Type the total revenue values into a blank Excel worksheet. Click on an empty cell nearby, click on the f_x icon, and insert the QUARTILE function. Move the window out of the way (if necessary) so that you can highlight the data for the **Array** text box. Then click in the **Quart** text box and enter a 1. Click on **OK** or hit ENTER. You should see the value of 79.775 (for $79.775) appear.

	D5		=	=QUARTILE(A1:A12,3)			
	A	B	C	D	E	F	G
1	98		$Q_1 =$	79.775			
2	74.1						
3	121.2		$Q_2 =$	84.45			
4	103.5						
5	79.4		$Q_3 =$	99.375			
6	89.3						
7	79.9						
8	80.2						
9	76.4						
10	109.7						
11	86.8						
12	82.1						

Figure 3.13 Using Excel's QUARTILE function in order to find the three quartiles. Quartile labels were typed in and subscripts were formatted. (To do this, highlight the numeral only, go to **Format>Cells**, click on the **Font** tab, and click on the checkbox next to **Subscript**.)

Repeat this for the 2nd and 3rd quartiles, entering a 2 and a 3 for each of them, respectively. You should see the values of 84.45 and 99.375 (for $84.45 million and $99.375 million) appear.

Since Excel uses a percentile method for calculating the three quartiles (which are really the 25th, 50th, and 75th percentiles), the values may differ slightly from those found by manual methods. But they still divide the data into four equal pieces (quarters) the same way.

Example 3-13 Refer to the data on revenues for the 12 top-grossing North American concert tours of all time given in Example 3-12. Use Excel to find the value of the 42nd percentile.

Solution: Type the total revenue values into a blank Excel worksheet (if you haven't done so already). Click on an empty cell nearby, click on the f_x icon, and insert the PERCENTILE function. Move the window out of the way (if necessary) so that you can highlight the data for the **Array** text box. Then click in the **K** text box and enter 0.42. Click on **OK** or hit ENTER. You should see the value of 81.378 (for $81.378 million) appear.

	C1	▼	=	=PERCENTILE(A1:A12,0.42)			
	A	B	C	D	E	F	G
1	98		81.378				
2	74.1						
3	121.2						
4	103.5						
5	79.4						
6	89.3						
7	79.9						
8	80.2						
9	76.4						
10	109.7						
11	86.8						
12	82.1						

Figure 3.14 Using Excel's PERCENTILE function to calculate the 42nd percentile.

Again, the value found by Excel may be slightly different from those found using manual methods, which are usually just approximations.

Percentile Rank

In order to find the percentile *rank* of a particular data value, use Excel's PERCENTRANK function. (Notice how the names of Excel's functions are really quite descriptive of what they do!)

Example 3-14 Refer to the data on revenues for the 12 top-grossing North American concert tours of all time given in Example 3-12. Use Excel to find the percentile rank for the revenue of $89.3 million..

Solution: Type the total revenue values into a blank Excel worksheet (if you haven't done so already). Click on an empty cell nearby, click on the f_x icon, and insert the PERCENTRANK function. Move the window out of the way (if necessary) so that you can highlight the data for the **Array** text box. Then click in the **X** text box, which is for the value in question, and then click on the cell that contains 89.3 (or just type in 89.3). The third text box is optional; you can specify how many decimal places if you want. The default is 3 decimal places, which is usually just fine. Click on **OK** or hit ENTER. You should see the value of 0.636 appear. This means that the value of $89.3 million is at the 63.6^{th} (or 64th) percentile.

	C1	▼	= =PERCENTRANK(A1:A12,A6)					
	A	B	C	D	E	F	G	
1	98		0.636					
2	74.1							
3	121.2							
4	103.5							
5	79.4							
6	89.3							
7	79.9							
8	80.2							
9	76.4							
10	109.7							
11	86.8							
12	82.1							

Figure 3.15 Using Excel's PERCENTRANK function in order to find the percentile associated with the value 89.3 (found in cell A6). It returns the value as a decimal (instead of percentage).

Again, the value may differ from approximations found by hand. But it still means the same thing: about 64% of these 12 North American concert tours grossed less than $89.3 million.

Exercises

3.1 Nixon Corporation manufactures computer terminals. The following data are the numbers of computer terminals produced at the company for a sample of 10 days.

24	32	27	23	35	33	29	40	23	28

Use Excel's built-in functions in order to find the mean, median, mode(s), range, variance, and standard deviation. Also find them by using Excel's Data Analysis ToolPak.

3.2 Use Excel to find the mean, median, mode(s), range, variance and standard deviation for each of the columns of data given in Exercise 1.4 (the prices of the five items in 25 urban areas across the United States). (*Exercise 1.4 explains how to calculate the means.*)

3.3 Use Excel to calculate the mean, variance, and standard deviation for the following grouped data, which represents a sample. You may choose your method (either generate columns and use mathematical formulas or use copies of midpoints and Excel's built-in functions).

x	2-4	5-7	8-10	11-13	14-16
f	5	9	14	7	5

3.4 The following table gives the frequency distribution of the number of errors committed by a college baseball team in all of the 45 games that they played during the 2002-2003 season.

Number of Errors	Number of Games
0	11
1	14
2	9
3	7
4	3
5	1

Use Excel to find the mean, variance, and standard deviation. (*Hint:* The classes in this example are single-valued. These values of classes will then be the values of the midpoints used in the formulas or functions. Also, notice that this data is from a population.)

CHAPTER **4**

Probability

CHAPTER OUTLINE

4.1 Calculating Probability

4.2 Conditional Probabilities and Contingency Tables

4.1 CALCULATING PROBABILITY

Just as we had Excel calculate *relative frequencies* from a frequency distribution table in Chapter 2 (see Example 2-3 in Section 2.2), we can have it calculate *probabilities* from a frequency distribution table. They are really exactly the same thing!

Example 4-1   Ten of the 500 randomly selected cars manufactured at a certain auto factory are found to be lemons.   Set up frequency and relative frequency (= probability) distribution tables in Excel. Assuming that the lemons are manufactured randomly, what is the probability that the next car manufactured at this auto factory is a lemon?

Solution: Set up column A to classify the cars as "Good" or "Lemon."   In column B, enter the corresponding frequencies (counts) out of 500 for each category. In column C, have Excel calculate the relative frequencies (= probabilities) for each category by dividing each frequency by the total of 500.

C3	▼	=	=10/500		
	A	B	C		D
1	Car	Frequency	Relative Frequency (= probability)		
2	Good	490	0.98		
3	Lemon	10	0.02		

Figure 4.1 Using Excel to calculate relative frequencies for approximating probabilities.

The probability that the next car is a lemon is now easily seen to be 0.02.

4.2 CONDITIONAL PROBABILITIES AND CONTINGENCY TABLES

When data is collected that involves two variables or characteristics, a two-way classification table or "contingency table" can be set up in order to show the overall distribution and then calculate probabilities.

Excel can generate a contingency table from the raw data itself via its **PivotTable** feature.

Example 4-2 All 100 employees of a company were asked whether they are in favor or against paying high salaries to CEOs of U.S. companies. Enter the resulting data as gender/opinion pairs into an Excel spreadsheet and then sort them by gender (male, then female) and then by opinion (in favor, then against). Then use Excel to generate a contingency table showing the distribution of gender (male or female) and opinion (in favor or against).

Solution: Enter each gender/opinion pair into rows 1-100, using column A for gender and column B for opinion. Highlight cells A1-B100 and go to **Data>Sort**. This is necessary, instead of just clicking on one of the sorting icons, in order to sort data by more than one characteristic and maintain pairs. In the **Sort** window, next to where it says **Sort by**, make sure that Column A is selected. Click on the radio button next to **Descending**, since we want males then females (reverse alphabetical order). Just below this, where it says **Then by**, select Column B. Also click on **Descending**, since we want "in favor" to come before "against" (again in reverse alphabetical order).

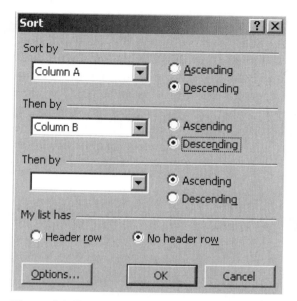

Figure 4.2 The sort options for sorting two-column, two-variable data by one characteristic and then another.

	A	B
1	M	In Favor
2	M	In Favor
3	M	In Favor
4	M	In Favor
5	M	In Favor
6	M	In Favor
7	M	In Favor
8	M	In Favor
9	M	In Favor
10	M	In Favor
11	M	In Favor
12	M	In Favor
13	M	In Favor
14	M	In Favor
15	M	In Favor
16	M	Against
17	M	Against
18	M	Against
19	M	Against
20	M	Against
21	M	Against
22	M	Against
23	M	Against
24	M	Against
25	M	Against

	A	B
26	M	Against
27	M	Against
28	M	Against
29	M	Against
30	M	Against
31	M	Against
32	M	Against
33	M	Against
34	M	Against
35	M	Against
36	M	Against
37	M	Against
38	M	Against
39	M	Against
40	M	Against
41	M	Against
42	M	Against
43	M	Against
44	M	Against
45	M	Against
46	M	Against
47	M	Against
48	M	Against
49	M	Against
50	M	Against

	A	B
51	M	Against
52	M	Against
53	M	Against
54	M	Against
55	M	Against
56	M	Against
57	M	Against
58	M	Against
59	M	Against
60	M	Against
61	F	In favor
62	F	In favor
63	F	In favor
64	F	In favor
65	F	Against
66	F	Against
67	F	Against
68	F	Against
69	F	Against
70	F	Against
71	F	Against
72	F	Against
73	F	Against
74	F	Against
75	F	Against

	A	B
76	F	Against
77	F	Against
78	F	Against
79	F	Against
80	F	Against
81	F	Against
82	F	Against
83	F	Against
84	F	Against
85	F	Against
86	F	Against
87	F	Against
88	F	Against
89	F	Against
90	F	Against
91	F	Against
92	F	Against
93	F	Against
94	F	Against
95	F	Against
96	F	Against
97	F	Against
98	F	Against
99	F	Against
100	F	Against

Figure 4.3 The data sorted by gender and opinion.

Now that the data is sorted, you could actually manually count the number of males in favor, males against, females in favor, and females against, so that you could set up a contingency table. But Excel will generate it for you!

First, we need to insert column headings. Click on a cell in row 1 and go to **Insert>Rows.** Now, type "Gender" in cell A1 and "Opinion" in cell B1. Highlight rows 1-101 of columns A and B (cells A1-B101) and go to **Data>PivotTable and PivotChart Report.** In steps 1 and 2 of the PivotChart and PivotTable Wizard, you can just click on **Next** (since you've highlighted the data, its source is known). But in step 3, *you must click on the* **Layout** button in order to set up the table. Drag the button for "Gender" to the ROW area, and the button for "Opinion" to the COLUMN area. Then also drag the button for "Opinion" to the DATA area. The window should then look like the one in the figure below.

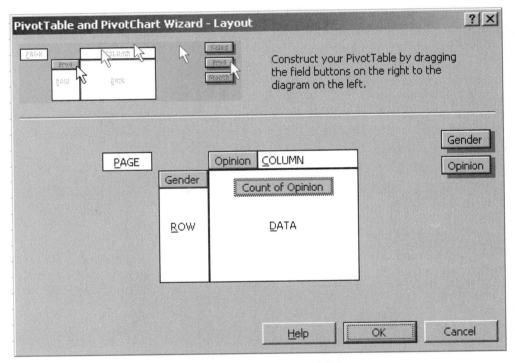

Figure 4.4 Set up the layout of your pivot table very carefully!

Click on **OK** to go back to step 3 and then click on **Finish**. You should see the two-way table with columns and rows in alphabetical order. If you want to change the order of the categories, click on the cell containing the category name and then drag its border to a new location. Here we want to have males listed first, so click on the cell with "M" in it, release it, and then drag its border up until you see a grayed line above the cell with "F" in it. Release it and make sure the category, with its counts, moves up a row. Drag the border of the cell with "Against" in it to the right so that it comes after the cell with "In favor" in it. Again, the counts will move as well. (Just make sure you click on the cell *and then release* before trying to drag its border.)

Count of Opinion	Opinion ▼		
Gender ▼	In Favor	Against	Grand Total
M	15	45	60
F	4	36	40
Grand Total	19	81	100

Figure 4.5 Excel's PivotTable shows a two-way classification of employee responses.

Now you can use Excel to calculate whatever probabilities (= relative frequencies) that you are interested in.

Example 4-3 Suppose one of the 100 employees surveyed in Example 4-2 is randomly selected. Use the contingency table in Figure 4.5 and Excel to calculate the following probabilities.

    a) P(female)
    b) P(in favor)
    c) P(in favor | female)
    d) P(female | in favor)
    e) P(female and in favor)
    f) P(female or in favor)

Solution: Enter the contingency table shown in Figure 4.5 into a blank Excel worksheet (unless you still have it from following along with the previous example).

    a) For the simple probability of female, you just need to divide the total number of females by the total number of employees. Click on an empty cell near the table, type "=" and then click on the cell with the total number of females in it (40), type "/" and click on cell with the total number of employees in it (100) and hit ENTER. You should see 0.4 appear.

    b) For the simple probability of in favor, you need to divide the total number of those in favor by the total number of employees. Click on an empty cell near the table, type "=" and then click on the cell with the total number of those in favor in it (19), type "/" and click on cell with the total number of employees in it (100) and hit ENTER. You should see 0.19 appear.

    c) For the conditional probability of in favor given female, you need to *focus only on the row of females*:

F		4	36	40

and divide the number of *females in favor* by the total number of *females*. Click on an empty cell near the table, type "=" and then click on the cell with the number of females in favor in it (4), type "/" and click on cell with the total number of females in it (40) and hit ENTER. You should see 0.1 appear.

d) For the conditional probability of female given in favor, you need to *focus only on the column of those in favor*:

In Favor
15
4
19

and divide the number of *females in favor* by the total number of those *in favor*. Click on an empty cell near the table, type "=" and then click on the cell with the number of females in favor in it (4), type "/" and click on cell with the total number of those in favor in it (19) and hit ENTER. You should see 0.210526 appear.

e) For the joint probability of female and in favor, you need to divide the number of *females in favor* by the total number of *employees*. Click on an empty cell near the table, type "=" and then click on the cell with the number of females in favor in it (4), type "/" and click on cell with the total number of employees in it (100) and hit ENTER. You should see 0.04 appear.

f) For the probability of female *or* in favor, you need to add the total number of females to the total number of those in favor, and then subtract the number of *females in favor*. Then divide this by the total number of employees. Click on an empty cell near the table, type "=(" and click on the cell with the total number of females in it (40), type "+" and click on the cell with the total number of those in favor in it (19), type "-" and click on the cell with the number of females in favor in it (4), type ")" and then "/" and click on cell with the total number of employees in it (100) and hit ENTER. You should see 0.55 appear.

B15	▼	**=** =(D6+B7-B6)/D7		
	A	B	C	D	E

	A	B	C	D	E
1					
2					
3	Count of Opinion	Opinion ▼			
4	Gender ▼	In Favor	Against	Grand Total	
5	M	15	45	60	
6	F	4	36	40	
7	Grand Total	19	81	100	
8					
9					
10	a) P(female)	0.4			
11	b) P(in favor)	0.19			
12	c) P(in favor | female)	0.1			
13	d) P(female | in favor)	0.210526			
14	e) P(female and in favor)	0.04			
15	f) P(female or in favor)	0.55			

Figure 4.6 Calculating probabilities from a contingency table in Excel.

Exercises

4.1 A sample of 2000 adults were asked whether or not they have ever shopped on the Internet. The following table gives a two-way classification of the responses.

	Have Shopped	Have Never Shopped
Male	300	900
Female	200	600

Enter the table into an Excel spreadsheet and have Excel calculate the totals (using the SUM function) of each column and row. Then have Excel calculate the following probabilities, if an adult is selected at random.

a) P(has never shopped on the Internet)

b) P(is male)

c) P(has never shopped on the Internet | male)

d) P(male | has never shopped on the internet)

e) P(is male and has never shopped on the Internet)

f) P(is male or has never shopped on the Internet)

4.2 Two thousand randomly selected adults were asked if they think they are financially better off than their parents. The following table gives the two-way classification of the responses based on the education levels of the persons included in the survey and whether they are financially better off, the same, or worse off than their parents.

	Less Than High School	High School	More Than High School
Better off	140	450	420
Same	60	250	110
Worse off	200	300	70

Enter the table into an Excel spreadsheet and have Excel calculate the totals (using the SUM function) of each column and row. Then have Excel calculate the following probabilities, if an adult is selected at random.

a) P(has more education than high school)

b) P(thinks he/she is financially better off than his/her parents)

c) P(has more education than high school | thinks he/she is financially better off than his/her parents)

d) P(thinks he/she is financially better off than his/her parents | has more education than high school)

e) P(thinks he/she is financially better off than his/her parents *and* has more education than high school)

f) P(thinks he/she is financially better off than his/her parents *or* has more education than high school)

CHAPTER **5**

Discrete Random Variables and their Probability Distributions

CHAPTER OUTLINE

5.1 Probability Distribution of a
    Discrete Random Variable
5.2 Mean and Standard Deviation of a
    Discrete Random Variable

5.3 Combinations
5.4 The Binomial Probability Distribution
5.5 The Hypergeometric Probability Distribution
5.6 The Poisson Probability Distribution

5.1 PROBABILITY DISTRIBUTION OF A DISCRETE RANDOM VARIABLE

As we saw in Chapter 4 (Section 4.1), Excel can be used to convert a frequency distribution table into a probability distribution table. Then a graphical presentation of this distribution can easily be generated.

Example 5-1   The table below gives the frequency distribution of the number of vehicles owned by all 2000 families living in a small town.

Number of Vehicles Owned	Frequency
0	30
1	470
2	850
3	490
4	160

Enter this into Excel as a probability distribution table (having Excel calculate relative frequencies in place of frequencies) and generate a graphical presentation of this distribution.

Solution: Type the column headings "Number of Vehicles Owned, x" and "Probability, $P(x)$" into cells A1 and B1. (Widen the columns as necessary.) Enter the values 0-4 into cells A2-A6. Click in cell B2, and type "=30/SUM(30,470,850,490,160)" and then copy this to cells B3-B6. Double-click on cell B3 so that you can edit it. Replace the 30 after the "=" sign with 470, so that it contains the formula "=470/SUM(30,470,850,490,160)." Hit ENTER. Edit cells B4-B6 similarly, so that they also contain the rest of the relative frequencies.

Highlight the table and click on the chart wizard icon: . In Step 1, select the Column chart type and click on **Next**. In Step 2, you need to click on the **Series** tab and remove the "Number of Vehicles Owned" series from the list, leaving only the "Probability" series. Then click in the **Category (X) axis labels** text box and highlight cells A2-A6. Click on **Next**. In Step 3, edit the chart title to read "Probability Distribution of the Number of Vehicles Owned by Families," enter "Number of Vehicles" for the x-axis label, and "Probability" for the y-axis label. Click on the **Legend** tab and remove the legend, and then click on **Finish**. Double-click on the bars, click on the **Options** tab, and lower the **Gap width** to 50 (or less). Resize the chart (by clicking on it and then dragging a corner) or move it (by clicking and dragging the whole thing) as necessary.

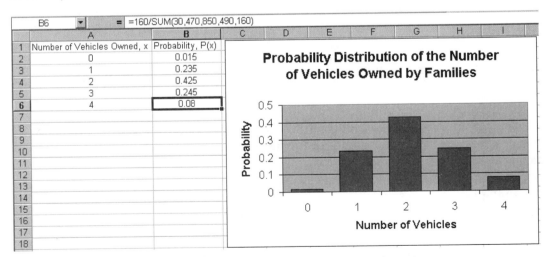

Figure 5.1 Using Excel to generate a probability distribution table and graph.

5.2 MEAN AND STANDARD DEVIATION OF A DISCRETE RANDOM VARIABLE

We will have Excel calculate the mean and standard deviation of a discrete random variable similarly to how we had it calculate the mean and standard deviation of grouped data in Chapter 3 (Section 3.3). The method we will use involves generating columns of values and using the mathematical formulas found in your textbook. (Unfortunately, Excel does not have one-step built-in functions for calculating these in this situation.)

Example 5-2   The following table lists the probability distribution of the number of breakdowns per week for a machine based on past data.

Breakdowns per week x	Probability $P(x)$
0	.15
1	.20
2	.35
3	.30

Use Excel to find the mean number of breakdowns per week for this machine. Also use it to calculate the standard deviation of the number of breakdowns.

Solution: Enter the probability distribution into a blank Excel worksheet. Include column headings of x and $P(x)$, using cells A1-B5. In cell C1, type "$xP(x)$" for the column heading for the products that will be used for calculating the mean. In cell C2, type "=" and click on cell A2, then type "*" and click on cell B2, and hit ENTER. Then copy this down to cells C3-C5. In cell C7, insert the SUM function (or type "=sum(" without the quotes), highlight cells C2-C5 and hit ENTER. You should see the mean value of 1.8 appear.

C6	▼	=	=SUM(C2:C5)		
	A	B	C	D	E
1	x	P(x)	xP(x)		
2	0	0.15	0		
3	1	0.2	0.2		
4	2	0.35	0.7		
5	3	0.3	0.9		
6			1.8		

Figure 5.2 Using Excel to find the mean of a probability distribution.

Now type a column label of "$xxP(x)$" or "$x^2P(x)$" in cell D1 for the products that will be used for calculating the standard deviation. Click on cell D2, type "=" and click on cell A2, then type "^2*" and click on cell B2. Then hit ENTER. Copy this down to cells D3-D5. Then copy cell C6 over to cell D6 to show the sum of that column, 4.3. Now, click on an empty cell nearby, say cell E6. Here we will finish calculating the standard deviation. Type "=SQRT(" and click on cell D6, then type "-" and click on cell C6, then type "^2)" and hit ENTER. You should see the standard deviation of 1.029563 appear.

	E6	▼	=	=SQRT(D6-C6^2)	
	A	B	C	D	E
1	x	P(x)	xP(x)	xxP(x)	
2	0	0.15	0	0	
3	1	0.2	0.2	0.2	
4	2	0.35	0.7	1.4	
5	3	0.3	0.9	2.7	
6			1.8	4.3	1.029563

Figure 5.3 Using Excel to find the standard deviation of a probability distribution.

5.3 FACTORIALS, PERMUTATIONS AND COMBINATIONS

Note: The functions used in the section are not on the list of **Statistical** functions in the **Paste Function** window. You have to scroll through **All** functions in order to find them.

Factorials

The value of the factorial of a number, $n! = n \times (n-1) \times \ldots \times 2 \times 1$, can be found with Excel's FACT function. In particular, this function can be used to find the number of arrangements of n items.

Example 5-3 How many ways can 10 books be arranged on a shelf?

Solution: The number of arrangements of 10 items is 10!. Click on a blank cell in an Excel worksheet and insert the FACT function. Type 10 and click on **OK** or hit ENTER and you should see the result of 3,628,800 appear.

	A1	▼	=	=FACT(10)
	A	B	C	D
1	3628800			

Figure 5.4 Excel's FACT function finds the number of arrangements of 10 books.

Permutations

The number of arrangements (i.e. permutations) of n items taken r at a time can be found using Excel's PERMUT function. This function requires two inputs:
1) the total number of items, n = **Number**
2) the number of items chosen per arrangement, r = **Number_chosen.**

Example 5-4 How many ways are there to arrange 7 out of 10 books on a shelf?

Solution: Click on a blank cell in an Excel worksheet and insert the PERMUT function. (*Note:* Remember that you can always just type the function name in after an "=" sign and followed by an open parenthesis. The two inputs then need to be separated with a comma.)

For the first input, type 10, for the total number of books. For the second input, type 7, for the number of books in the arrangement. You should see the value of 604,800 appear.

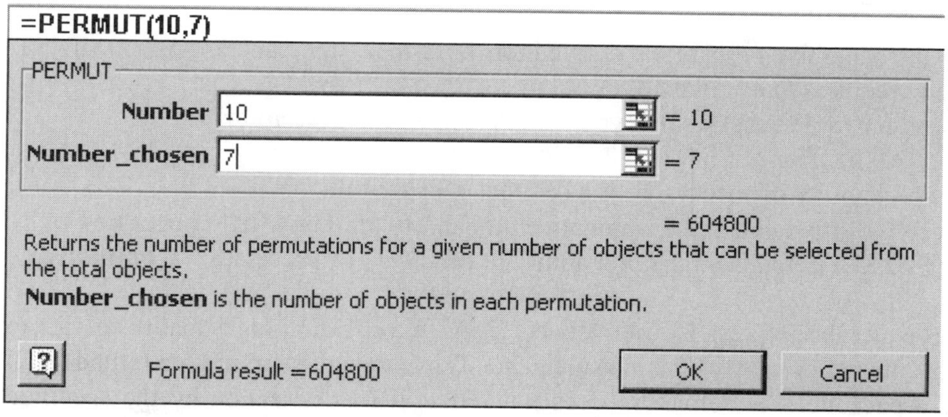

```
=PERMUT(10,7)
```

PERMUT

Number	10	= 10
Number_chosen	7|	= 7

= 604800

Returns the number of permutations for a given number of objects that can be selected from the total objects.

Number_chosen is the number of objects in each permutation.

Formula result =604800   [OK]   [Cancel]

Figure 5.5 Excel's PERMUT function finds the number of arrangements of 7 books out of 10.

Combinations

Perhaps most useful in probability and statistics is the combinations formula. Excel has this formula as a built-in function called COMBIN. This function requires two inputs:
3) the total number of items, n = **Number**
4) the number of items chosen per combination, x = **Number_chosen.**

Example 5-5 An ice cream parlor has six flavors of ice cream. Kristen wants to buy two flavors of ice cream. If she randomly selects two flavors out of six, how many possible combinations are there?

Solution: Click on a blank cell in an Excel worksheet and insert the COMBIN function. (*Note:* Remember that you can always just type the function name in after an "=" sign and followed by an open parenthesis. The two inputs then need to be separated with a comma.)

For the first input, type 6, for the total number of flavors of ice cream. For the second input, type 2, for the number of flavors chosen. You should see the value of 15 appear.

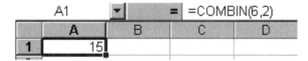

Figure 5.6 Excel's COMBIN function finds the number of two-flavor choices out of six.

5.4 THE BINOMIAL PROBABILITY DISTRIBUTION

Excel has the binomial probability distribution built-in as the function BINOMDIST. This function requires *four* inputs:
1) the number of successes that you are interested in, $x =$ **Number_s**
2) the total number of trials, $n =$ **Trials**
3) the probability of success, p, as a decimal = **Probability_s**
4) 1 (TRUE) if you want the cumulative probability of *at most* ($\le$) x successes, or 0 (FALSE) if you want the probability of *exactly* (=) x successes = **Cumulative**.

Example 5-6 At the Express House Delivery Service, providing high-quality service to customers is the top priority of the management. The company guarantees a refund of all charges if a package it is delivering does not arrive at its destination by the specified time. It is known from past data that despite all efforts, 2% of the packages mailed through this company do not arrive at their destinations within the specified time. Suppose a corporation mails 10 packages through Express House Delivery Service on a certain day. Use Excel to find
  a) the probability that exactly 1 of these 10 packages will not arrive at its destination within the specified time, and
  b) the probability that at most 1 of these 10 packages will not arrive at its destination within the specified time.

Solution: Click on a blank cell in an Excel worksheet. For part a), the first input is 1, for 1 of the 10 packages, the second input is 10, the total number of packages, the third input is .02, the probability that a package does not arrive by the specified time, and the fourth is 0 (for FALSE), since we want the probability for "exactly" 1. Click on **OK** and you should see the probability of 0.16675 appear. See Figure 5.7 below.

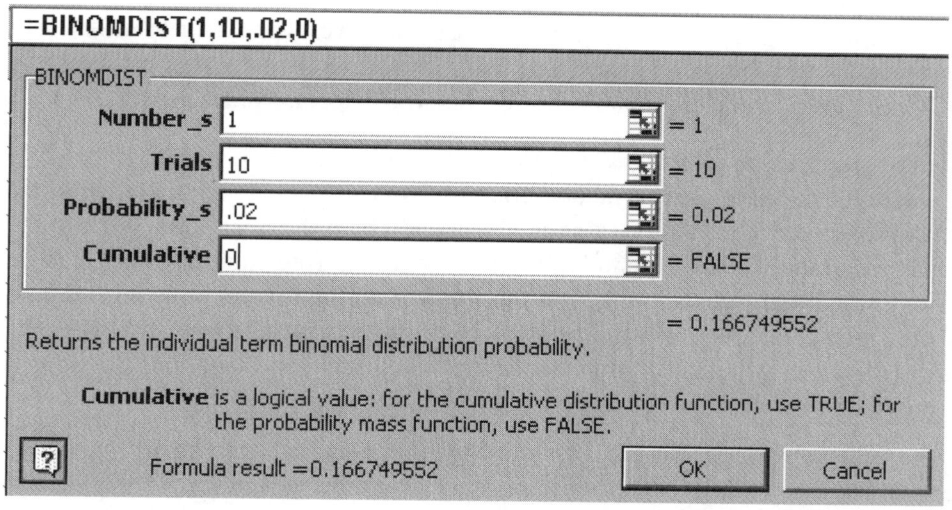

```
=BINOMDIST(1,10,.02,0)
```

BINOMDIST

Number_s 1   = 1
Trials 10   = 10
Probability_s .02   = 0.02
Cumulative 0| = FALSE

= 0.166749552

Returns the individual term binomial distribution probability.

Cumulative is a logical value: for the cumulative distribution function, use TRUE; for the probability mass function, use FALSE.

Formula result =0.166749552   OK   Cancel

Figure 5.7 Go to **Insert>Function** in order to insert Excel's BINOMDIST function, and fill in the four inputs in order to calculate a binomial probability.

For part b, the inputs are all the same except the last one, which needs to be a 1 (for TRUE), since we want the probability of "at most" (less than or equal to) 1. If you want to, you can just double-click on the cell and edit this input instead of reinserting the function and starting over. When you hit ENTER, you should see the probability of 0.983822 appear.

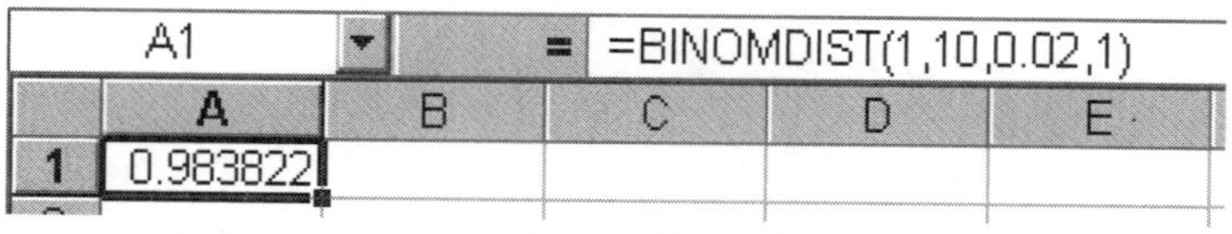

A1		=	=BINOMDIST(1,10,0.02,1)

	A	B	C	D	E
1	0.983822				

Figure 5.8 Excel's BINOMDIST function can be typed in and/or edited instead of inserted.

Excel's BINOMDIST function and chart wizard are also wonderful tools for viewing an entire binomial distribution.

Example 5-7 According to an Allstate Survey, 56% of Baby Boomers have car loans and are making payments on these loans (*USA TODAY*, October 28, 2002). Assume that this result holds true for the current population of all Baby Boomers. Let x denote the number in a random sample of three Baby Boomers who are making payments on their car loans. Have Excel generate the probability distribution of x and graph it.

Solution: Enter headings of "*x*" and "P(*x*)" into cells A1 and B1. Enter the possible values of *x*: 0, 1, 2, and 3, into cells A2-A5. Click on cell B2 and insert the BINOMDIST function. For the first input, **Number_s**, move the window out of the way, if necessary, and click on cell A2. For the second input, **Trials**, type a 3. For the third input, **Probability_s**, enter .56, and for the fourth input, **Cumulative**, type 0 (for FALSE). Hit ENTER and you should see the value of 0.085184 appear. Now copy this cell down to cells B3-B5, and you'll have the probabilities for the other success values as well!

For the graph, click on a blank cell and then click on the chart wizard icon: 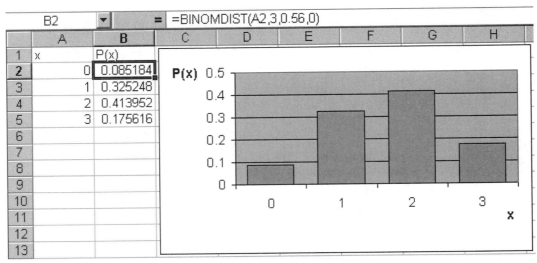. In Step 1, select the Column chart type and click on **Next**. In Step 2, highlight cells B2-B5 (the probabilities only) for the **Data range**. Then click on the **Series** tab, click in the **Category (X) axis labels** text box, and highlight cells A2-A5 (the values of *x*). Click on **Next**. In Step 3, edit the titles and click on the **Legend** tab and remove the legend, and then click on **Finish**. Double-click on the bars, click on the **Options** tab, and lower the **Gap width** to 50 (or less). Finally, resize the chart (by clicking on it and then dragging a corner) or move it (by clicking and dragging the whole thing) as necessary.

Figure 5.9 Using Excel's BINOMDIST function (with the FALSE option) and chart wizard to represent a binomial distribution.

Variations of the BINOMDIST function are sometimes required to answer specific probability questions. Remember that "at most" is the same as "equal or less" and "at least" is the same as "equal or more."

Example 5-8 According to a 2001 study of college students by Harvard University's School of Public Health, 19.3% of those included in the study abstained from drinking (*USA TODAY*, April 3, 2002). Suppose that of all current college students in the United

States, 20% abstain from drinking. A random sample of six college students is selected. Use Excel to answer the following.

a) Find the probability that exactly three college students in this sample abstain from drinking.
b) Find the probability that at most two college students in this sample abstain from drinking.
c) Find the probability that at least three college students in this sample abstain from drinking.
d) Find the probability that one to three college students in this sample abstain from drinking.

Solution: Using the BINOMDIST function for the various inequalities involved, with 0 (FALSE) for "=" and 1 (TRUE) for "$\leq$," the inputs and answers are as follows:
   a) $P(x = 3)$ is calculated by: =BINOMDIST(3, 6, .20, 0) and the result is: 0.08192.
   b) $P(x \leq 2)$ is calculated by: =BINOMDIST(2, 6, .20, 1) and the result is: 0.90112.
   c) $P(x \geq 3) = 1 - P(x < 3) = 1 - P(x \leq 2)$, so use:
      =1-BINOMDIST(2, 6, .20, 1) and the result is: 0.09888.
   d) $P(1 \leq x \leq 3) = P(x \leq 3) - P(x < 1) = P(x \leq 3) - P(x \leq 0)$, so use:
      = BINOMDIST(3, 6, .20, 1)-BINOMDIST(0, 6, .20, 1) and the result is: 0.720896.

5.5 THE HYPERGEOMETRIC PROBABILITY DISTRIBUTION

Excel has the hypergeometric probability distribution built in as the function HYPGEOMDIST. This function requires *four* inputs:
   1) the number of successes that you are interested in, x (in n trials) = **Sample_s**
   2) the number of trials, $n =$ **Number_sample**
   3) the number of successes in the population, $r =$ **Population_s**
   4) the total number of elements in the population, $N =$ **Number_pop.**

Example 5-9 Brown Manufacturing makes auto parts that are sold to auto dealers. Last week the company shipped 25 auto parts to a dealer. Later on, it found out that five of those parts were defective. By the time the company manager contacted the dealer, four auto parts from that shipment had already been sold. What is the probability that three of those four parts were good parts and one was defective?

Solution: Click on an empty cell in an Excel spreadsheet. Click on the f_x icon and insert the function HYPGEOMDIST. Identify and enter the four required inputs:
   1) The number of successes, i.e. good parts, that we are interested in, **Sample_s**, is 3.

2) The number of trials, **Number_sample**, is 4.
3) The number of successes, i.e. good parts, in the population, **Population_s**, is 20.
4) The total number of elements in the population, **Number_pop**, is 25.
Hit ENTER or click on **OK** and you should see the probability of 0.450593 appear.

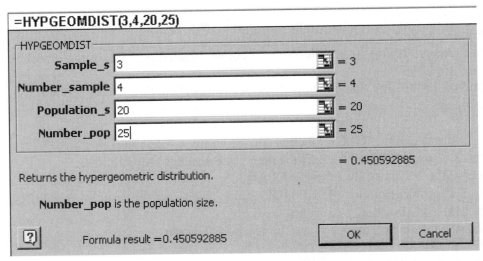

Figure 5.10 Fill in the four inputs required for the HYPGEOMDIST function in order to calculate a hypergeometric probability.

5.6 THE POISSON PROBABILITY DISTRIBUTION

Excel has the Poisson probability distribution built in as the function POISSON. This function requires *three* inputs:
1) the number of occurrences that you are interested in, $x = $ **X**
2) the mean number of occurrences, $\lambda = $ **Mean**
3) 1 (TRUE) if you want the cumulative probability of *at most* ($\leq$) x occurrences, or 0 (FALSE) if you want the probability of *exactly* (=) x occurrences = **Cumulative**.

Example 5-10 A washing machine in a laundromat breaks down an average of three times per month. Using Excel, find the probability that during the next month this machine will have
   a) exactly two breakdowns   b) at most one breakdown

Solution: Click on an empty cell in an Excel spreadsheet. Click on the f_x icon and insert the function POISSON. Identify and enter the three required inputs. For part a):
   1) The number of occurrences, i.e. breakdowns, that we are interested in, **X**, is 2.

2)   The mean number of occurrences, **Mean**, is 3.
3)   The logical value of **Cumulative** is FALSE, since we want the probability of *exactly* 2 occurrences. Enter 0.

Hit ENTER or click on **OK** and you should see the probability of 0.224042 appear.

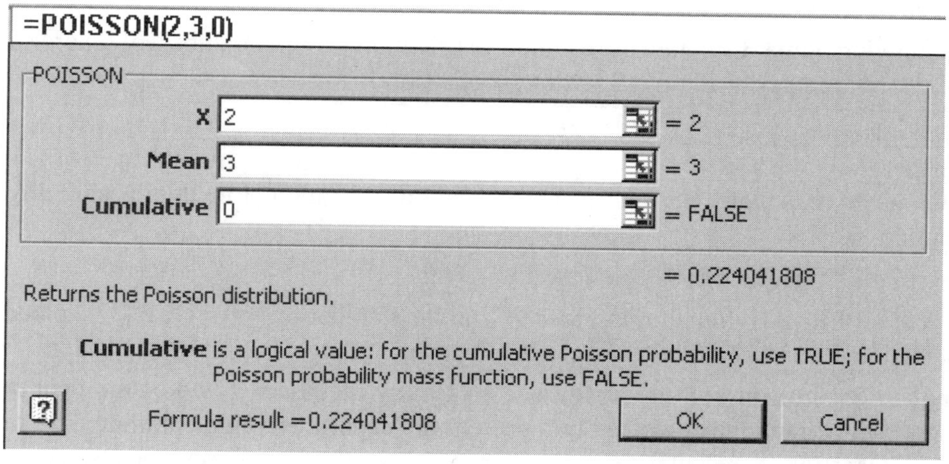

=POISSON(2,3,0)

POISSON		
X	2	= 2
Mean	3	= 3
Cumulative	0	= FALSE

= 0.224041808

Returns the Poisson distribution.

Cumulative is a logical value: for the cumulative Poisson probability, use TRUE; for the Poisson probability mass function, use FALSE.

Formula result =0.224041808   OK   Cancel

Figure 5.11 Fill in the four inputs required for the POISSON function in order to calculate a Poisson probability.

For part b), you can simply double-click on the same cell and edit the three inputs:
1)   The number of occurrences, i.e. breakdowns, that we are interested in, **X**, is 1.
2)   The mean number of occurrences, **Mean**, is still 3.
3)   The logical value of **Cumulative** is TRUE, since we want the cumulative probability of *at most* 1 occurrence. Change the 0 to a 1.

Hit ENTER and you should see the probability of 0.199148 appear.

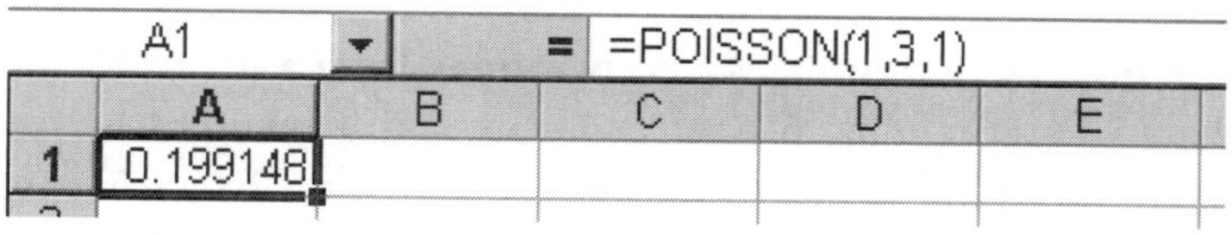

A1		=	=POISSON(1,3,1)	
	A	B	C	D	E
1	0.199148				

Figure 5.12 Excel's POISSON function can be typed in and/or edited instead of inserted.

Note that Excel's built-in probability distributions will allow you to be slightly more accurate in your calculations than using values from tables that have been rounded off. (Your answers may differ slightly from the answers in your textbook as a result.)

Exercises

5.1 One of the most profitable items at A1's Auto Security Shop is the remote starting system. Let *x* be the number of such systems installed on a given day at this shop. The following table lists the frequency distribution of *x* for the past 80 days.

x	1	2	3	4	5
f	8	20	24	16	12

Use Excel to convert this frequency distribution table to a probability distribution table for the number of remote starting systems installed on a given day. Then have Excel generate a graph of the probability distribution.

5.2 The following table lists the probability distribution of the number of exercise machines sold per day at Elmo's Sporting Goods store.

Machines sold per day	4	5	6	7	8	9	10
Probability	.08	.11	.14	.19	.20	.16	.12

Use Excel to calculate the mean and standard deviation for this probability distribution.

5.3 Use Excel's FACT function to calculate 11!.

5.4 Use Excel's PERMUT function in order to find the number of ways 1st, 2nd, 3rd place can be assigned to three of the eight women in an age group division of a triathlon.

5.5 An English department at a university has 16 faculty members. Two of the faculty members will be randomly selected to represent the department on a committee. Use Excel's COMBIN to calculate the number of ways the department can select 2 faculty members from 16.

5.6 In a poll of 12- to 18-year-old females conducted by Harris Interactive for the Gillette Company, 40% of the young females said that they expected the United States to have a female president within 10 years (*USA TODAY*, October 1, 2002). Assume that this result is true for the current population of all 12- to 18-year-old females. Suppose a random sample of 16 females from this age group is selected. Using Excel's BINOMDIST function, find the probability that the number of young females in this sample who expect a female president within 10 years is

a. at least 9   b. at most 5   c. 6 to 9.

5.7 An Internal Revenue Service inspector is to select 3 corporations from a list of 15 for tax audit purposes. Of the 15 corporations, 6 earned profits and 9 incurred losses during the year for which the tax returns are to be audited. If the IRS inspector decides to select three corporations randomly, use Excel's HYPGEOMDIST function to find the probability that the number of corporations in these three that incurred losses during the year for which the tax returns are to be audited is

a. exactly 2   b. none   c. at most 1.

5.6 A large proportion of small businesses in the United States fail during the first few years of operation. On average, 1.6 businesses file for bankruptcy per day in a large city. Using Excel's POISSON function, find the probability that the number of businesses that will file for bankruptcy on a given day in this city is

a. exactly 3   b. 2 to 3   c. more than 3   d. less than 3.

CHAPTER 6

Continuous Random Variables and the Normal Distribution

CHAPTER OUTLINE

6.1 THE STANDARD NORMAL DISTRIBUTION

The *standard* normal distribution is a normal distribution that has a mean equal to 0 and a standard deviation of 1. Probabilities and z-scores associated with this distribution can be found with Excel's NORMSDIST and NORMSINV functions, respectively. The "S" in the middle of each function name is for "Standard."

First, we'll look at finding probabilities for the standard normal distribution. (In Section 6.4, we'll calculate z-scores.) Your textbook has a table of probabilities associated with various z-scores that can be used, but Excel can calculate probabilities for *any* z-score with its built-in NORMSDIST function. This function calculates the *cumulative* probability for any z-score, i.e., the area under the curve from that z-score *on down* (to the left). So if you want the probability *between* two z-scores, you have to use the function for each one and then subtract, greater minus smaller. (Your answer, which is a probability or area, should always be positive.)

Example 6-1 Find the area under the standard normal curve between $z = 0$ and $z = 1.95$.

Solution: Click on a blank cell in an Excel worksheet. Type the following:
=NORMSDIST(1.95)-NORMSDIST(0).
Hit ENTER, and you should see the probability of 0.474412 appear.

The other option is to *insert* the NORMSDIST function, type 1.95 into the text box, click **OK**, hit F2 to edit the cell, type a minus sign after the last parenthesis, insert NORMSDIST again, type 0, and then click **OK** again.

Example 6-2 Use Excel to find the following areas under the standard normal curve.
a) Area to the right of $z = 2.32$
b) Area to the left of $z = -1.54$

Solution: For part a), you need to subtract the cumulative probability (area to the left) from 1 to get the complementary probability (area to the right). Type the following into a blank cell: =1-NORMSDIST(2.32).

You could, instead, just type "=1-" and then insert the function and enter 2.32. Or you could find the cumulative probability in one cell, and then in another cell type "=1-" and then click on the cell with the cumulative probability, and hit ENTER. Either way you do it, you should see the probability of 0.01017 appear.

For part b), just insert the NORMSDIST function (or type it in with "=" before it and "(" after it) and enter -1.54. This should give you the desired cumulative probability (area to the left) of 0.06178.

6.2 STANDARDIZING A NORMAL DISTRIBUTION

If you want to find the z-score that corresponds to some x-value from a *non*standard normal distribution (where the mean is not 0 or the standard deviation is not 1), then you can use Excel's STANDARDIZE function. This function has three inputs:
1) the x-value to be standardized
2) the mean of the normal distribution
3) the standard deviation of the normal distribution.

Example 6-3 Let x be a continuous random variable that has a normal distribution with a mean of 50 and a standard deviation of 10. Convert the following x-values to z-scores.
a) $x = 55$ b) $x = 35$

Solution: Click on a blank cell in an Excel worksheet, and insert the STANDARDIZE function. For part a), enter 55 into the first text box, 50 into the second, and 10 into the third. (You can hit TAB to go from one text box to the next, if you wish.) Click **OK** or hit ENTER, and you should see the z-score of 0.5 appear.

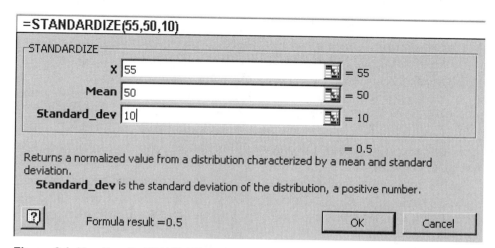

Figure 6.1 Use Excel's STANDARDIZE function if you want to convert an x-value from a normal distribution with given mean and standard deviation to a z-score.

For part b), you can simply edit the first input (double-click on the cell or click on it and hit F2).   Change the 55 to a 35 and hit ENTER. You should see a z-score of –1.5 appear.

A1	▼	=	=STANDARDIZE(35,50,10)		
	A	B	C	D	E
1	-1.5				

Figure 6.2 Excel's STANDARDIZE function can be typed in and/or edited instead of inserted.

In order to find *probabilities* associated with a nonstandard normal distribution (when the mean is not 0 or the standard deviation is not 1), you can either first standardize and use the NORMSDIST function, or you can simply use the NORMDIST function which takes into account the different mean and standard deviation. (Notice the absence of the "S" in the middle of the function name, since this is used for a normal distribution that is *not* standard.)

The NORMDIST function requires four inputs:
1) the x-value at the edge of the desired cumulative probability
2) the mean of the distribution
3) the standard deviation of the distribution
4) 1 (for TRUE) since we *always* want the cumulative normal probability.

(Note: This last input is different from the discrete distributions like the binomial distribution, where the FALSE option is sometimes desirable.)

Example 6-4 Let x be a continuous random variable that is normally distributed with a mean of 25 and a standard deviation of 4. Find the area
a) between $x = 25$ and $x = 32$ b) between $x = 18$ and $x = 34$.

Solution:

a) Click on a blank cell in an Excel worksheet. Type the following:
=NORMDIST(32,25,4,1)-NORMDIST(25,25,4,1).
Hit ENTER and you should see the probability of 0.459941 appear.

The other option is to *insert* the NORMDIST function, type 32 into the first text box, 25 into the second, 4 into the third, and 1 into the fourth text box, click **OK**, hit F2 to edit the cell, type a minus sign after the last parenthesis, insert NORMDIST again, enter 25, 25, 4, and 1, and then click **OK** again. (You can hit TAB to go from one text box to the next, if you wish.)

b) Double-click on the cell from part a) and edit the functions so that the first inputs are 34 and 18 instead of 32 and 25, respectively. Hit ENTER and you should see the probability of 0.947716 appear.

A1	▼	=	=NORMDIST(34,25,4,1)-NORMDIST(18,25,4,1)					
A	B	C	D	E	F	G	H
1 0.947716								

Figure 6.3 Using the difference of two applications of Excel's NORMDIST function to find a probability between two values.

Example 6-5 Let x be a continuous random variable that has a normal distribution with a mean of 80 and a standard deviation of 12. Find the area under the normal distribution curve to the left of 27.

Solution: This is a direct application of the NORMDIST function, since we want the area *to the left*. Insert this function into a cell and enter the four inputs as shown in the figure below. Click **OK** and you should see a value of 5.0159E-06 appear. This is Excel's scientific notation, and means 5.0159×10^{-6} or 0.0000050159. Pretty small! That's because 27 is more than 4 standard deviations ($4 \times 12 = 48$) below the mean (80). So the area below *that* is almost 0.

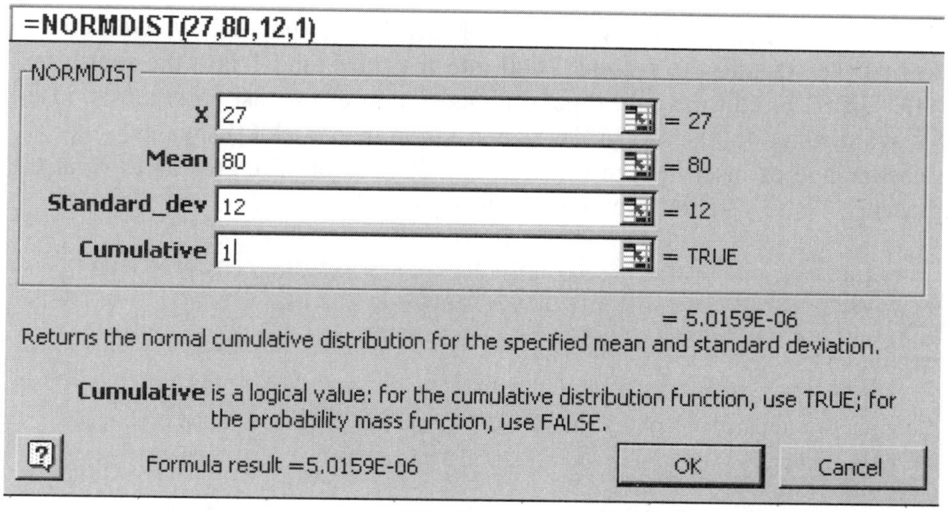

```
=NORMDIST(27,80,12,1)
```

NORMDIST

X	27	= 27
Mean	80	= 80
Standard_dev	12	= 12
Cumulative	1	= TRUE

= 5.0159E-06

Returns the normal cumulative distribution for the specified mean and standard deviation.

Cumulative is a logical value: for the cumulative distribution function, use TRUE; for the probability mass function, use FALSE.

Formula result = 5.0159E-06   OK   Cancel

Figure 6.4 Excel gives the result in scientific notation when the probability is extremely small.

6.3 APPLICATIONS OF THE NORMAL DISTRIBUTION

Now let's look at a couple of applications of Excel's NORMDIST function.

Example 6-6   According to *Automotive Lease Guide*, the Porsche 911 sports car is among the vehicles that hold their value best. A Porsche 911 (with a price of $87,500 for a new car) is expected to command a price of $48,125 after three years (*The Wall Street Journal*, August 6, 2002). Suppose the prices of all three-year-old Porsche 911 sports cars have a normal distribution with a mean price of $48,125 and a standard deviation of $1600. Find the probability that a randomly selected three-year-old Porsche 911 will sell for a price between $46,000 and $49,000.

Solution: Note that we have a normal distribution with a mean of 48,125 and a standard deviation of 1600. So to find the probability, we need to use the NORMDIST function. Those values will be the 2nd and 3rd inputs. Since we want the probability for *between* two values, we need to apply the function twice, once with the upper value and once with the lower value as the first input, and then subtract. This can be done simply by typing in the following:

=NORMDIST(49000,48125,1600,1)-NORMDIST(46000,48125,1600,1).

You should see the resulting probability of 0.615699, or approximately 61.6%. Note that you must *not* type commas within a number. Commas must *only* be used to *separate* numbers.

Of course, the other option is to *insert* the NORMDIST function, type 49000 into the first text box, 48125 into the second, 1600 into the third, and 1 into the fourth text box, click **OK**, hit F2 to edit the cell, type a minus sign after the last parenthesis, insert NORMDIST again, enter 46000, 48125, 1600, and 1, and then click **OK** again.

The advantage of inserting the function is that you get prompted as to what the input(s) represent, in case you don't remember what goes where.

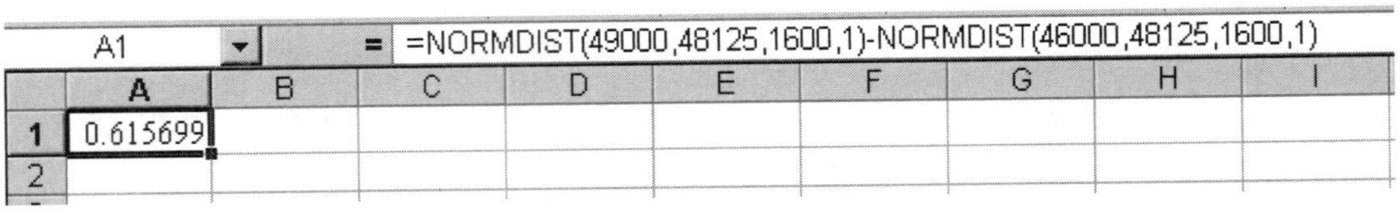

A1		=	=NORMDIST(49000,48125,1600,1)-NORMDIST(46000,48125,1600,1)					
A	B	C	D	E	F	G	H	I
1 0.615699								
2								

Figure 6.5 Subtract two applications of the NORMDIST function in order to find the probability between two values.

Example 6-7 A racing car is one of the many toys manufactured by Mack Corporation. The assembly times for this toy follow a normal distribution with a mean of 55 minutes and a standard deviation of 4 minutes. The company closes at 5 P.M. every day. If one worker starts to assemble a racing car at 4 P.M., what is the probability that she will finish this job before the company closes for the day?

Solution: Identify the mean and standard deviation: 55 and 4 minutes, respectively. These are the 2nd and 3rd inputs of Excel's NORMDIST function. The 1st input is the "cutoff" for the probability desired. In this case, we want the probability that she will finish in 1 hour (60 minutes) or less. So the first input is 60. Always enter 1 for the last input. Hit ENTER or click on **OK** and you will get the cumulative probability of 60 *or less*. This value, 0.89435 (approximately 89.4%), is exactly what we want!

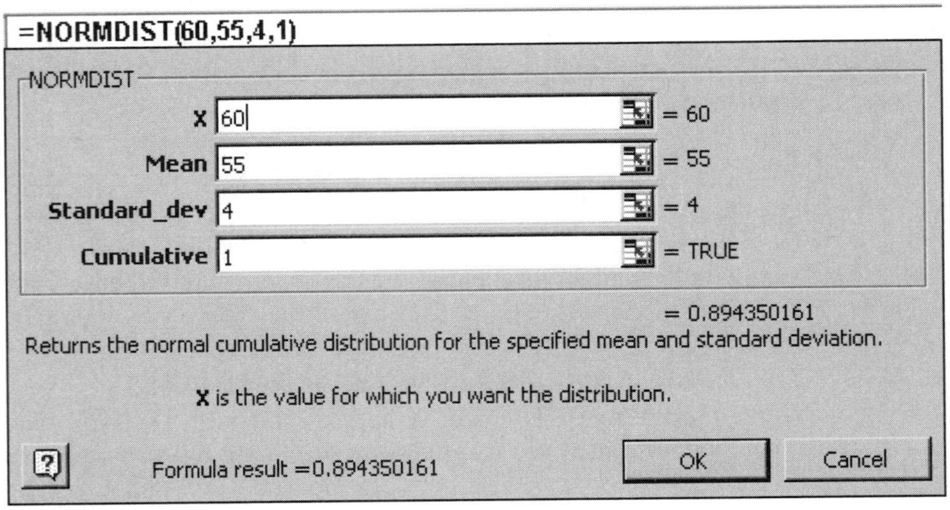

=NORMDIST(60,55,4,1)

NORMDIST

X 60 = 60
Mean 55 = 55
Standard_dev 4 = 4
Cumulative 1 = TRUE

= 0.894350161

Returns the normal cumulative distribution for the specified mean and standard deviation.

X is the value for which you want the distribution.

Formula result =0.894350161   OK   Cancel

Figure 6.6 Using Excel's NORMDIST function to find a cumulative probability.

6.4 DETERMINING Z AND X VALUES WHEN AN AREA UNDER THE NORMAL CURVE IS KNOWN

In order to go from a cumulative probability *back* to a z-score, use the function NORMSINV (with an "S"). In order to go from a cumulative probability back to an x-value in a nonstandard normal distribution, you need to use the function NORMINV (with no "S").   The key is to make sure that you first know the *cumulative* area or probability, i.e. the area or probability *below* or *to the left of* the desired value.

Example 6-8 Find the value of z such that the area under the standard normal curve in the right tail is .0050.

Solution: Since the area to the *right* of z is .0050, the area to the *left* of z, the *cumulative* area, must be 1-.0050 = .9950. Insert the NORMSINV function and enter .9950 (or you can simply enter 1-.0050).   Click on **OK** and you should see the z-score of 2.575835 appear.

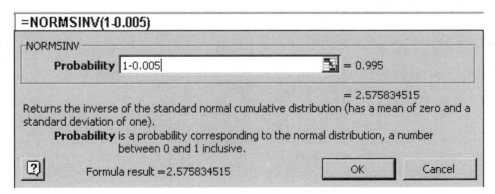

Figure 6.7 Use Excel's NORMSINV function to go "backwards" from a cumulative area (to the left of a z-score in the standard normal distribution) to the z-score.

The first input of the NORMINV function is also the cumulative probability (or area to the the left) of the desired value, but there are two more inputs required: the mean and standard deviation of the nonstandard normal distribution.

Example 6-9 It is known that the life of a calculator manufactured by Intal Corporation has a normal distribution with a mean of 54 months and a standard deviation of 8 months. What should the warranty period be to replace a malfunctioning calculator if the company does not want to replace more than 1% of all the calculators sold?

Solution: Note that we have a normal distribution with a mean of 54 and standard deviation of 8. Since we want to find a "cutoff" value for the 1% proportion (or probability), we need to use the NORMINV function. Now determine the cumulative probability – the probability *below* the cutoff that we're looking for. If the 1% are being replaced, it's because they are not lasting long enough. They are the ones on the low end of the distribution. So this is our cumulative probability: 0.01.

Insert the NORMINV function and enter 0.01, 54, and 8. Hit ENTER and you should see the value of 35.38926, for approximately 35.4 months. This is what the warranty period should be.

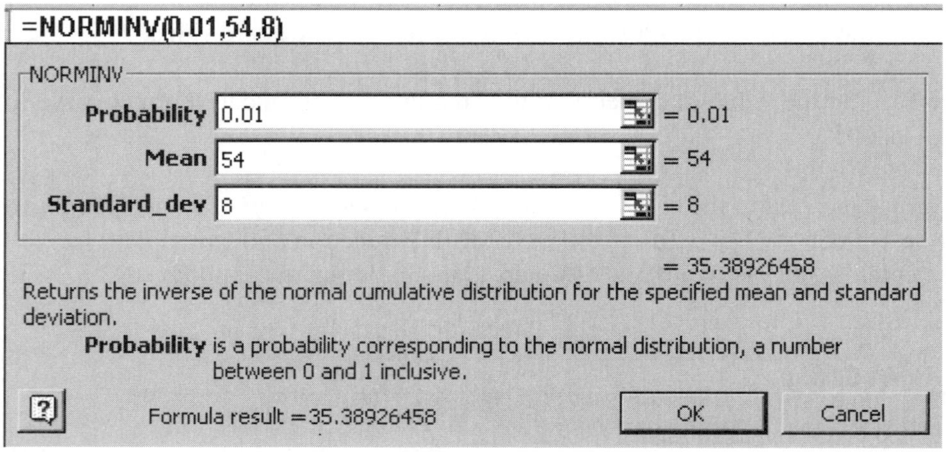

```
=NORMINV(0.01,54,8)
```

NORMINV		
Probability	0.01	= 0.01
Mean	54	= 54
Standard_dev	8	= 8

= 35.38926458

Returns the inverse of the normal cumulative distribution for the specified mean and standard deviation.

Probability is a probability corresponding to the normal distribution, a number between 0 and 1 inclusive.

Formula result = 35.38926458   OK   Cancel

Figure 6.8 Use Excel's NORMINV function to find a "cutoff" value for a known cumulative proportion or probability in a nonstandard normal distribution.

Example 6-10 Almost all high school students who intend to go to college take the SAT test. In 2002, the mean SAT score (in verbal and mathematics) of all students was 1020. Debbie is planning to take this test soon. Suppose the SAT scores of all students who take this test with Debbie will have a normal distribution with a mean of 1020 and a standard deviation of 153. What should her score be on this test so that only 10% of all examinees score higher than she does?

Solution: Here we have a normal distribution with a mean of 1020 and standard deviation of 153. Since we want to find a "cutoff" value for the 10% proportion (or probability), we need to use the NORMINV function. But this is *not* the cumulative probability . This is the percentage *above* the value we're looking for. The cumulative probability is 100%-10% = 90% or 0.90. Notice that this is talking about the 90th percentile!

Insert the NORMINV function and enter 0.90, 1020, and 153. Hit ENTER and you should see the value of 1216.077. She needs to make a score that is at least 1216.

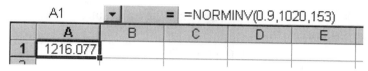

Figure 6.9 Using Excel's NORMINV function to find the 90th percentile in a normal distribution.

Exercises

6.1 Use Excel's NORMSDIST function to find the area under the standard normal curve

    a. between $z = 0$ and $z = -1.85$ b. between $z = 1.15$ and $z = 2.37$

    c. from $z = -1.53$ to $z = -2.88$ d. from $z = -1.67$ to $z = 2.44$

    e. to the right of $z = 1.56$ f. to the left of $z = -1.97$.

6.2 Use Excel's STANDARDIZE function in order to find the z-score for each of the following x-values for a normal distribution with $\mu = 30$ and $\sigma = 5$.

    a. $x = 39$ b. $x = 17$ c. $x = 22$ d. $x = 42$

6.3 Use Excel's NORMDIST function to find the following areas under a normal distribution curve with $\mu = 20$ and $\sigma = 4$.

    a.   Area between $x = 20$ and $x = 27$.

    b.   Area from $x = 23$ to $x = 25$.

    c.   Area between $x = 9.5$ and $x = 17$.

6.4 Use Excel's NORMDIST function in order to determine the area under a normal distribution curve with $\mu = 55$ and $\sigma = 7$

    a. to the right of $x = 58$ b. to the left of $x = 43$.

6.5 The U.S. Bureau of Labor Statistics conducts periodic surveys to collect information on the labor market. According to one such survey, the average earnings of workers in retail trade were $10 per hour in August 2002 (*Bureau of Labor Statistics News, September 18,2002*). Assume that the hourly earnings of such workers in August 2002 had a normal distrbution with a mean of $10 and a standard deviation of $1.10. Use Excel's NORMDIST function to find the probability that the hourly earnings of a randomly selected retail trade worker in August 2002 were

    a. more than $12   b. between $8.50 and $10.80.

6.6 The transmission on a model of a specific car has a warranty for 40,000 miles. It is known that the life of such a transmission has a normal distribution with a mean of 72,000 miles and a standard deviation of 12,000 miles. Use Excel's NORMDIST function to find the following.

    a.   What percentage of the transmissions will fail before the end of the warranty period?

    b.   What percentage of the transmissions will be good for more than 100,000 miles?

6.7 Use Excel's NORMSINV function in order to determine the value of z so that the area under the standard normal curve

    a. in the right tail is .0500   b. in the left tail is .0250

    c. in the left tail is .0100   d. in the right tail is .0050

6.8 Fast Auto Service provides oil and lube service for cars. It is known that the mean time taken for oil and lube service at this garage is 15 minutes per car and the standard devation is 2.4 minutes. The management wants to promote the business by guaranteeing a maximum waiting time for its customers. If a customer's car is not serviced within that period, the customer will receive a 50% discount on the charges. The company wants to limit this discount to at most 5% of the customers. Assume that the times taken for oil and lube service for all cars have a normal distribution. Use Excel's NORMINV function to find what the maximum guaranteed waiting time should be.

CHAPTER 7

Sampling Distributions

CHAPTER OUTLINE

7.1 APPLICATIONS OF THE SAMPLING DISTRIBUTION OF $\bar{x}$

According to the Central Limit Theorem, the sampling distribution of $\bar{x}$ is approximately normal for large samples ($n \geq 30$) as well as for samples (of any size) taken from a normally distributed population. Therefore, we can use Excel's NORMDIST function in order to calculate probabilities involving sample means in these cases. The input for the mean of the sampling distribution is the same as the mean of the population, μ, and the input for the standard deviation is the standard deviation of the population divided by the square root of the sample size. This can be entered as $\sigma/\text{SQRT}(n)$, with the known values of σ and n inserted.

Example 7-1 Assume that the weights of all packages of a certain brand of cookies are normally distributed with a mean of 32 ounces and a standard deviation of .3 ounce. Find the probability that the mean weight, $\bar{x}$, of a random sample of 20 packages of this brand of cookies will be between 31.8 and 31.9 ounces.

Solution: Although the sample size is small ($n < 30$), the shape of the sampling distribution of $\bar{x}$ is normal because the population is normally distributed. In order to calculate the probability of $\bar{x}$ being *between* two values, we need to use Excel's

NORMDIST function to calculate the two associated cumulative probabilities and subtract the upper one from the lower one.

Click on a blank cell in an Excel spreadsheet and insert the NORMDIST function. The first input is the upper limit of the interval, 31.9. The second input is the mean of the sampling distribution, which is the same as the mean of the population, 32. The third input is the standard deviation of the sampling distribution, which is the standard deviation of the population divided by the square root of the sample size. Enter .3/SQRT(20). And we *always* want the fourth input to be 1 for this function. Click on **OK** and you should see the upper cumulative probability of 0.068019 appear. This is *not* your final answer, of course; you need to subtract the lower cumulative probability from it.

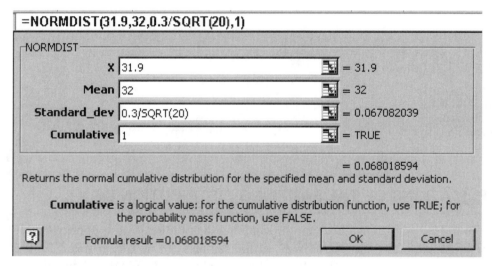

Figure 7.1 Using NORMDIST to calculate probabilities involving the sampling distribution of $\bar{x}$.

Hit F2 in order to edit the cell. The cursor should be at the end of the NORMDIST function that you just entered. Type a minus sign and then insert the NORMDIST function again, this time entering the lower limit of the interval, 31.8 for the first input. Enter the others the same as before. Click on **OK** and you should see the desired probability of 0.066584 appear.

Remember that you can either insert functions or type them (if you remember which input is which).

A1	▼	=	=NORMDIST(31.9,32,0.3/SQRT(20),1)-NORMDIST(31.8,32,0.3/SQRT(20),1)						
A	B	C	D	E	F	G	H	I	J
1 0.066584									
2									

Figure 7.2 The required calculation involving the NORMDIST function can be typed in and/or edited as well as inserted.

Example 7-2 According to the College Board's report, the average tuition and fees at four-year private colleges and universities in the United States was $18,273 for the academic year 2002-2003 (*The Hartford Courant*, October 22, 2002). Suppose that the shape of the distribution of the 2002-2003 tuition and fees at all four-year private colleges in the United States was unknown, but its mean was $18,273 and the standard deviation was $2100. Let $\bar{x}$ be the mean tuition and fees for 2002-2003 for a random sample of 49 four-year private U.S. colleges.

a) What is the probability that the 2002-2003 mean tuition and fees for this sample was within $550 of the population mean?
b) What is the probability that the 2002-2003 mean tuition and fees for this sample was lower than the population mean by $400 or more?

Solution: Although the shape of the distribution of the population (2002-2003 tuition and fees at all four-year private colleges in the United States) is unknown, the sampling distribution of $\bar{x}$ is approximately normal because the sample is large ($n \geq 30$). Remember that when the sample is large, the Central Limit Theorem applies. So again we can use Excel's NORMDIST function, entering the interval boundaries and the mean and standard deviation of the sampling distribution.

Note that the population mean, μ, is 18,273, the population standard deviation is 2100, and the sample size is 49. So the mean and standard deviation for the sampling distribution of $\bar{x}$ that we want to enter into the function as the 2nd and 3rd inputs are 18273 and 2100/SQRT(49), respectively. And the 4th input is always 1.

For part a), we are interested in the probability that $\bar{x}$ is between 18273-550 and 18273+550. So we need the difference of two applications of the NORMDIST function as in the previous example, the first with the upper value entered as the 1st input, and the second with the lower value entered as the 1st input. Either insert the function, enter the inputs, hit ENTER, type the minus sign, insert the function again, enter the inputs, and hit ENTER again, or type it all in as shown in Figure 7.3.

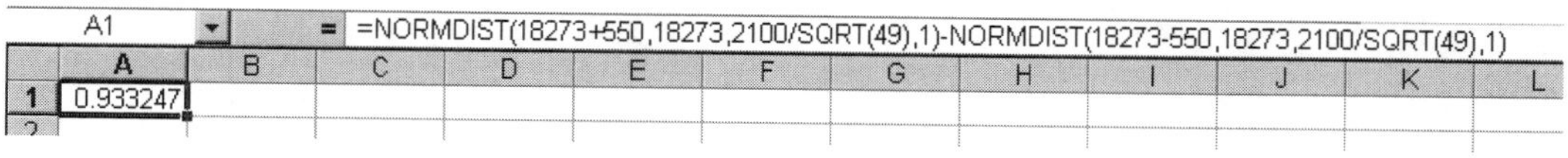

A1			=	=NORMDIST(18273+550,18273,2100/SQRT(49),1)-NORMDIST(18273-550,18273,2100/SQRT(49),1)
A	B	C	D	E F G H I J K L
1	0.933247			

Figure 7.3 Using Excel's NORMDIST function to find the probability that $\bar{x}$ is within $550 of the population mean.

For part b), we want the cumulative probability that $\bar{x}$ will be at least $400 lower than the population mean, i.e. that it will be $18273-400 or less. This is a direct application of the NORMDIST function. Insert the function and enter the inputs as shown in Figure 7.4. Click on **OK** and you should see the probability of 0.091211 appear.

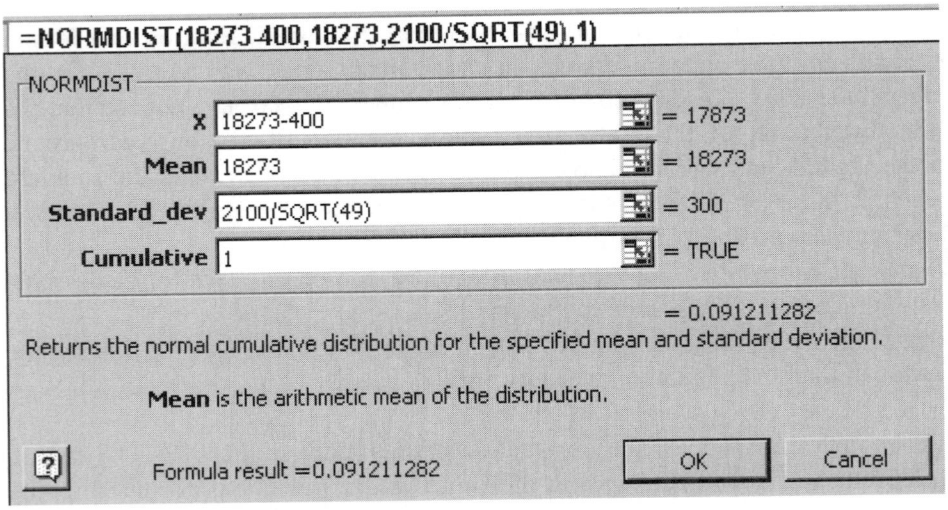

```
=NORMDIST(18273-400,18273,2100/SQRT(49),1)
```

Figure 7.4 Using Excel's NORMDIST function to find the probability that $\bar{x}$ was lower than the population mean by $400 or more.

7.2 APPLICATIONS OF THE SAMPLING DISTRIBUTION OF $\hat{p}$

According to the Central Limit Theorem, the sampling distribution of $\hat{p}$ is approximately normal for sufficiently large sample size. In the case of proportion, the sample size is considered to be sufficiently large if np and nq are both greater than 5. Therefore, we can use Excel's NORMDIST function in order to calculate probabilities involving sample proportions in these cases. Or we can first standardize the values with the STANDARDIZE function to obtain z-scores, and then use Excel's NORMSDIST function (with an "S").

The input for mean of the sampling distribution of $\hat{p}$ is the population proportion, p, and the input for the standard deviation is given by $\sqrt{\dfrac{pq}{n}}$. This can be entered as SQRT(pq/n), with the known values of p, q, and n inserted (or as SQRT($p*(1-p)/n$), with the known values of p and n inserted).

Example 7-3 Maureen Webster, who is running for mayor in a large city, claims that she is favored by 53% of all eligible voters of that city. Assume that this claim is true. What is the probability that in a random sample of 400 registered voters taken from this city, less than 49% will favor Maureen Webster?

Solution: First check that $np > 5$ and $nq > 5$ so that we can use the normal distribution. The assumption is that $p = 53\%$ or .53, so $q = 1-.53 = .47$. The sample size, n is 400.

Click on a blank cell in an Excel worksheet, and type "=.53*400" and hit ENTER, and you should see 212 appear. This is definitely greater than 5. Double-click on the cell and change the .53 to .47 and hit ENTER, and you should see 188 appear, also greater than 5. So the sampling distribution is normal.

Click on another empty cell in the worksheet, and insert the NORMDIST function. We want the probability that $\hat{p}$ is less than 49%, so enter .49 for the first input. The second input is p, so enter .53. For the third input, enter SQRT(.53*.47/400), which gives the value of the standard deviation using the formula noted above. Finally, enter 1 for the last input, as usual. In this case, the cumulative probability is exactly what we want. Hit ENTER to get the result of 0.05448.

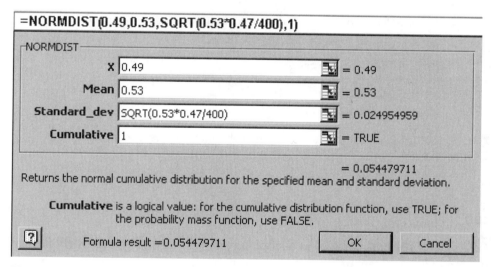

Figure 7.5 Using Excel's NORMDIST function to calculate a probability involving a sample proportion.

The other option is to first find the z-score associated with a sample proportion of $\hat{p} = .49$, and then use the NORMSDIST function. The STANDARDIZE function is used for this, and the inputs for mean and standard deviation are the same as above.

Using this method, we see the associated z-score is –1.60289. Then the probability, of course, comes out the same.

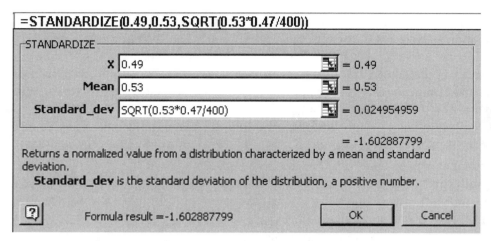

Figure 7.6 Using STANDARDIZE in order to find the z-score associated with a sample proportion.

	C1	▼	=	=NORMSDIST(A1)	
	A	B	C	D	E
1	-1.60289		0.05448		
2					

Figure 7.7 Using NORMSDIST with a reference to the cell containing the z-score to find the associated cumulative probability.

Exercises

7.1 The GPAs of all students enrolled at a large university have an approximately normal distribution with a mean of 3.02 and a standard deviation of .29. Use Excel's NORMDIST function (and the Central Limit Theorem) to find the probability that the mean GPA of a random sample of 20 students selected from this university is

    a. 3.10 or higher   b. 2.90 or lower   c. 2.95 to 3.11

7.2 According to CardWeb.com, the average credit card debt per household was $8367 in 2001 (*USA TODAY*, April 29, 2002). Assume that the probability distribution of all such current debts is skewed to the right with a mean of $8367 and a standard deviation of $2400. Use Excel's NORMDIST function (and the Central Limit Theorem) to find the probability that the mean of a random sample of 225 such debts is

    a.   between $8100 and $8500

    b.   within $200 of the population mean

    c.   greater than the population mean by $300 or more.

7.3 The National Institute of Child Health and Human Development conducted a study of bullying in U.S. schools in which over 15,000 students in sixth through tenth grades were surveyed. Of students who have been bullied, 61.6% said that they were victimized because of their looks or speech (*U.S. News & World Report*, May 7, 2001). Assume that this percentage is true for all current sixth- through tenth-grade students who have been bullied. Let $\hat{p}$ be the proportion in a random sample of 200 such students who will say that they are victimized because of their looks or speech. Find the probability that the value of $\hat{p}$ is

a. between .60 and .66   b. greater than .64

Use Excel's STANDARDIZE function to find associated z-scores first, and then use NORMSDIST to find the probabilities.

CHAPTER 8

Estimation of the Mean and Proportion

CHAPTER OUTLINE

8.1 INTERVAL ESTIMATION OF A POPULATION MEAN: LARGE SAMPLES

When the sample size is large ($n \geq 30$), the maximum error of estimate for a confidence interval for μ can be found with Excel's CONFIDENCE function. This function has three inputs:

1) the confidence level subtracted from 100%, entered as a decimal, $\alpha =$ **Alpha**
2) the standard deviation of the population, $\sigma =$ **Standard_dev**
   (use s, the sample standard deviation, if σ is not known)
3) the sample size, $n =$ **Size**.

This maximum error of estimate can then be subtracted from and added to the sample mean in order to construct the interval.

Example 8-1 A publishing company has just published a new college textbook. Before the company decides the price at which to sell this textbook, it wants to know the average price of all such textbooks in the market. The research department at the company took a sample of 36 such textbooks and collected information on their prices. This information produced a mean price of $70.50 for this sample. It is known that the standard deviation of the prices of all such textbooks is $4.50.

    a) What is the point estimate of the mean price of all such college textbooks?
    b) What is the maximum error of estimate for a 95% confidence interval?
    c) Construct a 90% confidence interval for the mean price of all such college textbooks.

Solution:
    a) The point estimate is the sample mean, 70.50. Enter this value into cell A1 of a new Excel spreadsheet.
    b) For the maximum error of estimate for a 95% confidence interval, click on cell A2, click on the f_x icon, and select the CONFIDENCE function. The first input is 100% – 95% = 5% entered as a decimal. You can either enter .05 or just 1-.95. The second input is the standard deviation of the population. (Use the sample standard deviation, if the population one is not known). Enter 4.50. (This function will automatically divide that by the square root of n for the standard deviation of the sampling distribution.) The third input is the sample size. Enter 36. Click on **OK** and you should see the maximum error of estimate is 1.469971, or ±$1.47.

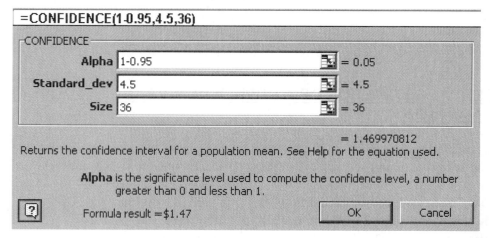

Figure 8.1 Using Excel's CONFIDENCE function to find the maximum error of estimate for a 95% confidence interval. (This is also called the "margin of error.")

    c) For the 90% confidence interval, we first need the maximum error of estimate. Copy cell A2 (from part b) to cell A3, double-click on it in order to edit it, and change the first input from 1-0.95 to 1-0.90. Hit ENTER and you should see the maximum error of estimate change to 1.23364, or $1.23. Now click on cell A4, type "=" and click on cell A1 (where the sample mean is), type "-" and click on cell A3 (where the maximum error of estimate is), and hit ENTER. You should see the lower limit of the confidence interval appear: 69.26636 or $69.27. Click on cell A5, type "=" and click on cell A1, type "+" and click on cell A3, and hit ENTER. You should see the upper limit of the confidence interval appear: 71.73364 or $71.73. Format these to two-decimal-place currency if you wish (go to **Format>Cells**, click on the **Number** tab and select **Currency**).

	A	B	C	D	E	F	G
	A5		= =A1+A3				
1	$70.50	= sample mean					
2	$1.47	= maximum error of estimate for 95% confidence interval					
3	$1.23	= maximum error of estimate for 90% confidence interval					
4	$69.27	= lower limit of 90% confidence interval					
5	$71.73	= upper limit of 90% confidence interval					

Figure 8.2 Use Excel's CONFIDENCE function with addition and subtraction to get the upper and lower limits of the confidence interval. (Answers may differ slightly from those found using values from a normal distribution table.)

Example 8-2 According to a report by the Consumer Federation of America, National Credit Union Foundation, and the Credit Union National Association, households with negative assets carried an average of $15,528 in debt in 2002 (CBS.MarketWatch.com, May 14, 2002). Assume that this mean was based on a random sample of 400 households and that the standard deviation of debts for households in this sample was $4200. Make a 99% confidence interval for the 2002 mean debt for all such households.

Solution: Click on an empty cell in an Excel worksheet and insert the CONFIDENCE function. For a 99% confidence interval, $\alpha = 1\%$ or .01. This is the first input. The population standard deviation is unknown, but we can use the sample standard deviation of $4200 for an approximation. This is the second input. The sample size is 400. This is the third input. Hit ENTER and you should see that the maximum error of estimate is 540.9252 or $540.93. Click on another empty cell and type "=15528-" and click on the cell with the maximum error of estimate in it. Hit ENTER and you should see that the lower limit of the confidence interval is 14987.07. Click on another empty cell and type "=15528+" and click on the cell with the maximum error of estimate in it again. Hit ENTER and you should see that the upper limit of the confidence interval is 16068.93.

A1		▼	=	=CONFIDENCE(0.01,4200,400)	
A	B	C	D	E	F
1 540.9252		14987.07	to	16068.93	

Figure 8.3 Using Excel's CONFIDENCE function with addition and subtraction to get the upper and lower limits of the 99% confidence interval. (Answers may differ slightly from those found using values from a normal distribution table.)

8.2 INTERVAL ESTIMATION OF A POPULATION MEAN: SMALL SAMPLES

In order to construct a confidence interval when a small sample ($n < 30$) is drawn from a normally distributed population with unknown standard deviation, we will use Excel's TINV function to find values based on the t-distribution. This function requires two inputs:
  1) the confidence level subtracted from 100%, entered as a decimal = **Probability** (note that this is the area of the *two* tails in the t-distribution)
  2) one less than the sample size, $n - 1 =$ **Deg_freedom**.

Then we will calculate the maximum error of estimate using the formula $t\dfrac{s}{\sqrt{n}}$, which will be entered into Excel as $=t*s/\text{SQRT}(n)$, with the known values for (or cell references to) t, s, and n inserted. (You can*not* use Excel's CONFIDENCE function for small samples, since it uses values based on the normal distribution.)

Example 8-3 Dr. Moore wanted to estimate the mean cholesterol level for all adult men living in Hartford. He took a sample of 25 adult men from Hartford and found that the mean cholesterol level for this sample is 186 with a standard deviation of 12. Assume that the cholesterol levels for all adult men in Hartford are (approximately) normally distributed. Construct a 95% confidence interval for the population mean, μ.

Solution: There are three steps:
  1) Use TINV(α, $n-1$) to find the value of t associated with a confidence level of 95% ($\alpha = .05$) and sample size of 25 ($n-1 = 24$).

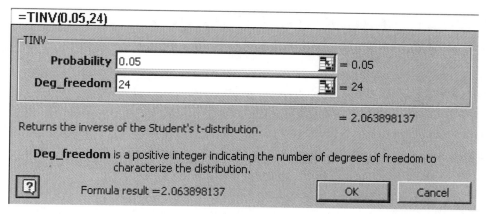

Figure 8.4 Using Excel's TINV function to obtain the value of t associated with a 95% confidence interval with a small sample size of 25.

2) Calculate the maximum error of estimate using the formula $=t*s/SQRT(n)$ using the value of t calculated in step 1, $s = 12$, and $n = 25$.

A2		=	=A1*12/SQRT(25)		
	A	B	C	D	E
1	2.063898				
2	4.953356				

Figure 8.5 Using the formula for the maximum error of estimate with a cell reference to the t-value calculated previously.

3) Add and subtract the value calculated in step 2 to and from the sample mean of 186 in order to get the upper and lower limits of the interval.

C3		=	=186+A2		
	A	B	**C**	D	E
1	2.063898				
2	4.953356				
3	181.0466	to	190.9534		

Figure 8.6 Adding the maximum error of estimate to (and subtracting it from) the mean value of 186 in order to get the upper (and lower) limit of the confidence interval.

8.3 INTERVAL ESTIMATION OF A POPULATION PROPORTION: LARGE SAMPLES

In order to construct a confidence interval for a population proportion, p, when we have a large sample, we can use Excel's CONFIDENCE function again. (The criteria for "large sample" are that $n\hat{p}$ and $n\hat{q}$ are both greater than 5.) In this case, we will enter SQRT(pq) for the second input, the standard deviation, approximating p with $\hat{p}$ and q with $\hat{q}$. (This function will then automatically divide that by the square root of n for an estimate of the standard deviation of the sampling distribution.)

Example 8-4 According to a 2002 survey by FindLaw.com, 20% of Americans needed legal advice during the past year to resolve such thorny issues as family trusts and landlord disputes (CBS.MarketWatch.com, August 6, 2002). Suppose a recent sample of 1000 adult Americans showed that 20% of them needed legal advice during the past year to resolve such family-related issues.
   a. What is the point estimate of the population proportion?
   b. What is the maximum error of estimate for a 95% confidence interval?
   c. Find a 99% confidence interval for the percentage of all adult Americans who needed legal advice during the past year to resolve such family-related issues.

Solution:
   a) The point estimate is the sample proportion, 20% or 0.20. Enter this value into cell A1 of a new Excel spreadsheet.
   b) For the maximum error of estimate for a 95% confidence interval, click on cell A2, click on the f_x icon, and select the CONFIDENCE function. The first input is 100% – 95% = 5% entered as a decimal. You can either enter .05 or just 1-.95. The second input is the standard deviation of the population, which we will approximate with the sample standard deviation. Note that $\hat{p}$ = .20, so $\hat{q}$ = 1-.20 = .80. So enter SQRT(.2*.8). The third input is the sample size. Enter 1000. Click on **OK** and you should see the maximum error of estimate is 0.024792, or ± 2.5%.

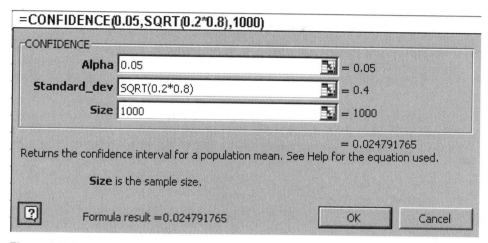

Figure 8.7 Using Excel's CONFIDENCE function to find the maximum error of estimate for a 95% confidence interval for *p*. (This is also called the "margin of error.")

    c)   For the 99% confidence interval, we first need the maximum error of estimate. Copy cell A2 (from part b) to cell A3, double-click on it in order to edit it, and change the first input from .05 to .01. Hit ENTER and you should see the maximum error of estimate change to .032582, or 3.3%. Now click on cell A4, type "=" and click on cell A1 (where the sample proportion is), type "-" and click on cell A3 (where the maximum error of estimate is), and hit ENTER. You should see the lower limit of the confidence interval appear: 0.167418 or 16.7%. Click on cell A5, type "=" and click on cell A1, type "+" and click on cell A3, and hit ENTER. You should see the upper limit of the confidence interval appear: 0.232582 or 23.3%. Format these to percentages with one decimal place if you wish (go to **Format>Cells**, click on the **Number** tab and select **Percentage**, and change the number of decimal places to 1).

	A5	▼	**=**	=A1+A3			
	A	**B**	**C**	**D**	**E**	**F**	**G**
1	20.0%	= sample proportion					
2	2.5%	= maximum error of estimate for 95% confidence interval					
3	3.3%	= maximum error of estimate for 99% confidence interval					
4	16.7%	= lower limit of 99% confidence interval					
5	23.3%	= upper limit of 99% confidence interval					

Figure 8.8 Use Excel's CONFIDENCE function with addition and subtraction to get the upper and lower limits of the confidence interval for *p*.

8.4 DETERMINING THE SAMPLE SIZE FOR THE ESTIMATION OF MEAN

In order to find the sample size required for estimating the mean with a given level of confidence and maximum error of estimate, you can either calculate it using the formula in your textbook, or you can use trial-and-error with the CONFIDENCE function.

For example, what would be the minimum required sample size if you want to be 99% confident that the sample mean is within two units of the population mean, given that $\sigma = 1.4$?

Try $n = 30$, and see what you get: CONFIDENCE(0.01, 1.4, 30) = 0.658. That's well within 2 units, so try a smaller one, say $n = 15$.

CONFIDENCE(0.01, 1.4, 15) = 0.931. Still that's less than 2. Try $n = 5$.

CONFIDENCE(0.01, 1.4, 5) = 1.612. That's closer to 2. What about $n = 4$?

CONFIDENCE(0.01, 1.4, 4) = 1.803. That's even closer. What about $n = 3$?

CONFIDENCE(0.01, 1.4, 3) = 2.082. That's not within 2 units anymore. Therefore, $n = 4$ is the minimum sample size required.

Example 8-5   An alumni association wants to estimate the mean debt of this year's college graduates. It is known that the population standard deviation of the debts of this year's college graduates is $11,800. How large a sample should be selected so that the estimate with a 99% confidence level is within $800 of the population mean?

Solution: Start by typing =CONFIDENCE(0.01, 11800, 30) into an empty cell in an Excel worksheet, since 1-.99 = .01 and the standard deviation is 11800. This sample size of 30 gives a maximum error of estimate of 5549.314, which is way too big. So we must have a much larger sample in order to get it down around 800. Try changing the "30" to "1000." This gives a value of 961.1695. This is much closer! Keep changing the sample size, n, in CONFIDENCE(0.01, 11800, n) in order to see where it changes from above 800 to below 800. You should find that $n = 1444$ gives a maximum error of estimate of 799.8644 whereas $n = 1443$ gives a maximum error of estimate of 800.1415. So to be within $800, the minimum sample size required is 1444.

8.5 DETERMINING THE SAMPLE SIZE FOR THE ESTIMATION OF PROPORTION

Just as with the mean, we can find the sample size required for estimating the proportion with a given level of confidence and maximum error of estimate either by calculating it using the formula in your textbook or by using trial-and-error with the CONFIDENCE function.   Again, we need to estimate the values of p and q so that we can enter

SQRT(pq) for the standard deviation. The most conservative estimate is to use $p = q =$ .50. Otherwise, we can use values of $\hat{p}$ and $\hat{q}$ from a preliminary sample.

Example 8-6 Lombard Electronics Company has just installed a new machine that makes a part that is used in clocks. The company wants to estimate the proportion of these parts produced by this machine that are defective. The company manager wants this estimate to be within .02 of the population proportion for a 95% confidence level. What is the most conservative estimate of the sample size that will limit the maximum error to within .02 of the population proportion?

Solution: Start by typing =CONFIDENCE(0.05, SQRT(.5*.5), 30) into an empty cell in an Excel worksheet, since 1-.95 = .05 and we're using $p = q = .50$ for the most conservative estimate. This sample size of 30 gives a maximum error of estimate of 0.178919, which is way too big. So we must have a much larger sample in order to get it down around 0.02. Try changing the "30" to "1000." This gives a value of 0.03099. This is much closer! Keep changing the sample size, n, in CONFIDENCE(0.05, SQRT(.5*.5), n) in order to see where it changes from above 0.02 to below 0.02. You should find that $n = 2401$ gives a maximum error of estimate of 0.0199996, whereas $n = 2400$ gives a maximum error of estimate of 0.020004. So to be within 0.02, the minimum sample size required is 2401.

Example 8-7 Consider Example 8-6 again. Suppose a preliminary sample of 200 parts produced by this machine showed that 7% of them are defective. How large a sample should the company select so that the 95% confidence interval for p is within .02 of the population proportion?

Solution: Start by typing =CONFIDENCE(0.05, SQRT(.07*.93), 30) into a an empty cell in an Excel worksheet, since 1-.95 = .01 and we're using $\hat{p} = .07$ and $\hat{q} = .93$ from the preliminary sample. This sample size of 30 gives a maximum error of estimate of 0.091301, which is still too big. We must have a much larger sample in order to get it down around 0.02. Again change the "30" to "1000." This gives a value of 0.015814. This is too small! So try a *smaller* sample, say $n = 500$. This gives a value of 0.022364, much closer. As before, keep changing the sample size, n, in CONFIDENCE(0.05, SQRT(.07*.93), n) in order to see where it changes from above 0.02 to below 0.02. You should find that $n = 626$ gives a maximum error of estimate of 0.019987, whereas $n = 625$ gives a maximum error of estimate of 0.020003. So to be within 0.02, the minimum sample size required is 626.

Exercises

8.1   According to *Money* magazine, the average net worth of U.S. households in 2002 was $355,000, (*Money*, Fall 2002). Assume that this mean is based on a random sample of 500 households and that the sample standard deviation is $125,000. Use Excel's CONFIDENCE function to construct a 99% confidence interval for the 2002 mean net worth of all U.S. households.

8.2   According to a 2002 survey by America Online, mothers with children under age 18 spent an average of 16.87 hours per week online (*USA TODAY*, May 7, 2002). Suppose that this mean is based on a random sample of 1000 such mothers and that the standard deviation for this sample is 3.2 hours per week.   Use Excel's CONFIDENCE function to construct a 95% confidence interval for the corresponding population mean for all such mothers.

8.3   A random sample of 16 airline passengers at the Bay City airport showed that the mean time spent waiting in line to check in at the ticket counters was 31 minutes with a standard deviation of 7 minutes. Use Excel's TINV function and the formula for the maximum error of estimate in order to construct a 99% confidence interval for the mean time spent waiting in line by all passengers at this airport. (Assume that such waiting times for all passengers are normally distributed.)

8.4   A random sample of 20 acres gave a mean yield of wheat equal to 41.2 bushels per acre with a standard deviation of 3 bushels. Assume that the yield of wheat per acre is normally distributed, and use Excel's TINV function and the formula for the maximum error of estimate in order to construct a 90% confidence interval for the population mean, μ.

8.5   In a Maritz poll of 1004 adult drivers conducted in July 2002, 36% said that they "often" or "sometimes" talk on their cell phones while driving (*USA TODAY*, October 23, 2002). Assume that these 1004 drivers make a random sample of all adult drivers in the United States. Use Excel's CONFIDENCE function in order to answer the following:

a.   What is the point estimate of the corresponding population proportion?
b.   What is the maximum error of estimate for a 95% confidence interval?
c.   Construct a 99% confidence interval for the proportion of all adult drivers in the United States who "often" or "sometimes" talk on their cell phones while driving.

8.6   A marketing researcher wants to find a 95% confidence interval for the mean amount that visitors to a theme park spend per person per day. She knows that the standard deviation of the amounts spent per person per day by all visitors to this park is $11. Use trial and error with Excel's CONFIDENCE function to determine how large a sample the researcher should select so that the estimate will be within $2 of the population mean.

8.7   Tony's Pizza guarantees all pizza deliveries within 30 minutes of the placement of orders. An agency wants to estimate the proportion of all pizzas delivered within

30 minutes by Tony's. Use trial and error with Excel's CONFIDENCE function to determine the most conservative estimate of the sample size that would limit the maximum error to within .02 of the population proportion for a 99% confidence interval.

8.8 Refer to exercise 8.7. Assume that a preliminary study has shown that 93% of all Tony's pizzas are delivered within 30 minutes. Use trial and error with Excel's CONFIDENCE function to determine how large the sample size should be so that the 99% confidence interval for the population proportion has a maximum error of .02.

CHAPTER 9

Hypothesis Tests about the Mean and Proportion

CHAPTER OUTLINE

9.1 Hypothesis Tests about μ for a Large Sample Using the p-value Approach

9.2 Hypothesis Tests about a Population Mean: Large Samples

9.3 Hypothesis Tests about a Population Mean: Small Samples

9.4 Hypothesis Tests about a Population Proportion: Large Samples

9.1　HYPOTHESIS TESTS ABOUT μ FOR A LARGE SAMPLE USING THE *P*-VALUE APPROACH

By definition, the *p-value* is the smallest significance level at which the null hypothesis is rejected. In this section, we will use Excel to set up a hypothesis test and calculate this value.

Example 9-1 The management of Priority Health Club claims that its members lose an average of 10 pounds or more within the first month after joining the club. A consumer agency that wanted to check this claim took a random sample of 36 members of this health club and found that they lost an average of 9.2 pounds within the first month of membership with a standard deviation of 2.4 pounds. Find the *p*-value for this test.

Solution: Notice that the claim to be tested is that $\mu \geq 10$. Also identify the sample statistics: $n = 36$, $\bar{x} = 9.2$, and $s = 2.4$.

Step 1. Enter the null and alternative hypotheses into an Excel spreadsheet. Note that the claim is the null hypothesis, since it includes equality. (To get the Greek letter, μ, first just type it in as an "m." Then double-click on the cell to edit it, highlight *just* the "m," go to **Format>Cells**, and select **Symbol** for the **Font**.)

Step 2. Note that this is a left-tailed test, since the alternate hypothesis contains the "<" symbol.

Step 3. Use Excel's NORMDIST function to calculate the p-value, which for a left-tailed test is precisely the cumulative probability associated with the sample statistic, $\bar{x}$. So the inputs required are 9.2 for **X**, 10 for **Mean**, 2.4/SQRT(36) for **Standard_dev**, and 1 for **Cumulative** (as always).

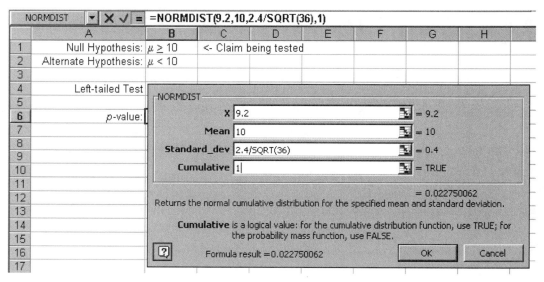

Figure 9.1 Using an Excel spreadsheet to organize a hypothesis test about a mean, and using the NORMDIST function to calculate the p-value.

The p-value in this case is 0.02275, which shows that for $\alpha = 0.05$, we would reject the null hypothesis (and thus the claim), but for $\alpha = 0.01$, we could *not*.

Example 9-2 At Canon Food Corporation, it took an average of 50 minutes for new workers to learn a food processing job. Recently the company installed a new food processing machine. The supervisor at the company wants to find if the mean time taken by new workers to learn the food processing procedure on this new machine is different from 50 minutes. A sample of 40 workers showed that it took, on average, 47 minutes for them to learn the food processing procedure on the new machine with a standard deviation of 7 minutes. Find the p-value for the test that the mean learning time for the food processing procedure on the new machine is different from 50 minutes.

Solution: Notice that the claim to be tested is that $\mu \neq 50$. This time, we'll enter the sample statistics into the Excel spreadsheet and use cell references to them when we calculate the p-value. Enter $n = 40$, $\bar{x} = 47$, and $s = 7$ into rows 6-8.

Step 1. Enter the null and alternate hypotheses and note that the claim is the alternate hypothesis since it does not include equality. (You can use "greater than or less than," "<>," for the not equal symbol if you wish.)

Step 2. Note that this is a two-tailed test.

Step 3. Calculate the p-value with the NORMDIST function. Since the sample mean of 47 is *below* the hypothesized mean of 50, use the NORMDIST function to find its cumulative area (area *below*) and multiply it by 2 as shown in the formula bar in the figure below. (If the sample statistic is *above* the hypothesized population parameter, then you need to subtract the cumulative area from 1 before doubling it.)

	B10	▼	=	=2*NORMDIST(B6,50,B7/SQRT(B8),1)		
	A	B	C	D	E	F
1	Null Hypothesis:	$\mu = 50$				
2	Alternate Hypothesis:	$\mu <> 50$	<- Claim being tested			
3						
4	Two-tailed Test					
5						
6	sample mean:	47				
7	sample st. dev.:	7				
8	sample size:	40				
9						
10	p-value:	0.006718				

Figure 9.2 Calculating the p-value for a two-tailed test when the sample statistic is below the hypothesized population parameter.

This p-value of 0.0067 is quite small, and would show rejection of the null hypothesis (and thus support for the claim) even at the highly statistically significant level of $\alpha = 0.01$.

9.2 HYPOTHESIS TESTS ABOUT A POPULATION MEAN: LARGE SAMPLES

Another way to make a decision about a claim is to calculate the "test statistic" and determine whether or not it falls into the "rejection region" for a predetermined level of significance, α. The examples in this section and the rest of this chapter will show this method *as well as* the calculation of the p-value.

Example 9-3 The TIV Telephone Company provides long-distance telephone service in an area. According to the company's records, the average length of all long-distance calls placed through this company in 1999 was 12.44 minutes. The company's management wanted to check if the mean length of the current long-distance calls is different from 12.44 minutes. A sample of 150 such calls placed through this company produced a mean length of 13.71 minutes with a standard deviation of 2.65 minutes. Using the 5% significance level, can you conclude that the mean length of all current long-distance calls is different from 12.44 minutes?

Solution: In order to set up the test, identify and enter the sample statistics into a blank Excel worksheet: $n = 150$, $\bar{x} = 13.71$, and $s = 2.65$. Note that the claim to be tested is that $\mu \neq 12.44$.

Step 1. Enter the null and alternate hypotheses and note that the claim is the alternate hypothesis since it does not include equality. (You can use "greater than or less than," "<>," for the not equal symbol if you wish.)

Step 2. Note that this is a two-tailed test.

Step 3. Determine the critical values corresponding to the significance level of $\alpha = .05$. Since this is a two-tailed test, the area of one tail is .05/2. To find the associated z-score, use NORMSINV(.05/2) to get the one on the left (the negative one). In this case of a two-tailed test, we want both the positive and negative values: ±1.959961.

Step 4. Calculate the value of the test statistic. We want the z-score associated with the sample statistics. In other words, we want to standardize our sample statistics. Insert the STANDARDIZE function into a blank cell, with inputs 13.71 for **X**, 12.44 for **Mean**, and 2.65/SQRT(150) for **Standard_dev**. (Remember that you can click on the cells containing these values instead of typing them in if you wish.) You should obtain a value of 5.869532.

Step 5. Since the test statistic of approximately 5.87 is well beyond the critical value of approximately 1.96, we reject the null hypothesis and conclude that evidence supports the claim (the alternate hypothesis).

Notice that this test statistic is very extreme. It is more than 5 standard deviations above the mean! The associated p-value can be found by typing in:

=2*(1-NORMSDIST(5.87))

which results in 4.385×10^{-9}, or 0.000000004385. This is so rare that the null hypothesis must be rejected at essentially *any* level of significance!

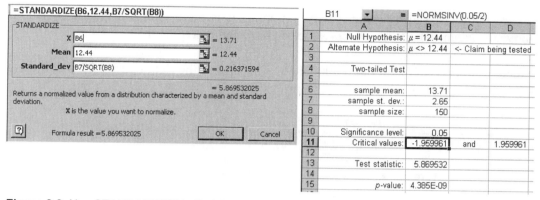

Figure 9.3 Use STANDARDIZE to find the test statistic corresponding to the sample statistics, and use NORMSINV to find the (negative) critical value for a given level of significance, α.

Example 9-4 According to a salary survey by National Association of Colleges and Employers, the average salary offered to computer science majors who graduated in May 2002 was $50,352 (*Journal of Accountancy*, September 2002). Suppose this result is true for all computer science majors who graduated in May 2002. A random sample of 200 computer science majors who graduated this year showed that they were offered a mean salary of $51,750 with a standard deviation of $5240. Using the 1% significance level, can you conclude that the mean salary of this year's computer science graduates is higher than $50,352?

Solution: In order to set up the test, identify and enter the sample statistics into a blank Excel worksheet: $n = 200$, $\bar{x} = 51750$, and $s = 5240$. Note that the claim to be tested is that $\mu > 50352$.

Step 1. Enter the null and alternate hypotheses and note that the claim is the alternate hypothesis since it does not include equality.

Step 2. Note that this is a right-tailed test, since the alternate hypothesis has the symbol ">" in it.

Step 3. Determine the critical value corresponding to the significance level of α = .01. Since this is a right-tailed test, the area of the right tail is .01. To find the associated z-score, use NORMSINV(.01) to get the negative z-score that has cumulative area of .01. In this case of a right-tailed test, we want the positive value: 2.326342.

Step 4. Calculate the value of the test statistic. We want the z-score associated with the sample statistics. In other words, we want to standardize our sample statistics. Insert the STANDARDIZE function into a blank cell, with inputs 51750 for **X**, 50352 for **Mean**, and 5240/SQRT(200) for **Standard_dev**. (Remember that you can click on the cells containing these values instead of typing them in if you wish.) You should obtain a value of 3.7730354.

Step 5. Since the test statistic of approximately 3.77 is well beyond the critical value of approximately 2.33, we reject the null hypothesis and conclude that evidence supports the claim (the alternate hypothesis).

This test statistic is also very extreme. It is more than 3 standard deviations above the mean. The associated p-value can be found by typing in:

=1-NORMSDIST(3.77)

which results in 8.066×10^{-5}, or 0.00008066. This area to the right of $z = 3.77$ is definitely less than $\alpha = 0.01$!

	B15	▼	=	=1-NORMSDIST(B13)		
	A	B	C	D	E	
1	Null Hypothesis: $\mu \leq 50352$					
2	Alternate Hypothesis: $\mu > 50352$		<- Claim being tested			
3						
4	Right-tailed Test					
5						
6	sample mean:	51750				
7	sample st. dev.:	5240				
8	sample size:	200				
9						
10	Significance level:	0.01				
11	Critical value:	-2.326342	--->	2.326342		
12						
13	Test statistic:	3.7730354				
14						
15	p-value:	8.066E-05				
16						

Figure 9.4 Setting up and calculating the critical value (=−NORMSINV(.01)), the test statistic (=STANDARDIZE(51750, 50352, 5240/SQRT(200)), and the p-value (see formula bar above) for this right-tailed hypothesis test. Reject H_0 and accept the claim.

Example 9-5 The mayor of a large city claims that the average net worth of families living in this city is at least $300,000. A random sample of 100 families selected from this city produced a mean net worth of $288,000 with a standard deviation of $80,000. Using the 2.5% significance level, test the mayor's claim.

Solution: In order to set up the test, identify and enter the sample statistics into a blank Excel worksheet: $n = 100$, $\bar{x} = 288000$, and $s = 80000$. Note that the claim to be tested is that $\mu \geq 300000$.

Step 1. Enter the null and alternate hypotheses and note that the claim is the null hypothesis since it *does* include equality.

Step 2. Note that this is a left-tailed test, since the alternate hypothesis has the symbol "<" in it.

Step 3. Determine the critical value corresponding to the significance level of α = .025. Since this is a left-tailed test, the area of the left tail is .025. To find the associated z-score, use NORMSINV(.025). This gives the desired negative z-score that has cumulative area of .025, approximately −1.96.

Step 4.   Calculate the value of the test statistic, i.e. standardize the sample statistics to find the associated z-score. Insert the STANDARDIZE function into a blank cell, click on the cell with the sample mean in it for **X**, enter the hypothesized population mean of 300000 for **Mean**, and enter 80000/SQRT(10) (or click on the references for the sample standard deviation and sample size) for **Standard_dev**. You should obtain a value of −1.5.

Step 5.   Since the test statistic of −1.5 is *not* beyond the critical value of −1.96, we can *not* reject the null hypothesis (claim) and therefore conclude that *there is not enough* evidence to conclude that the mayor's claim is false. It is probably true.

The associated p-value can be found by typing in:

=NORMSDIST(−1.5)

which results in 0.0668072. This area to the left of $z = −1.5$ is not less than $\alpha = 0.025$, which also shows the failure to reject the null hypothesis in this test.

B13		=	=STANDARDIZE(B6,300000,B7/SQRT(B8))		
	A	B	C	D	E
1	Null Hypothesis:	μ ≥ 300000	<- Claim being tested		
2	Alternate Hypothesis:	μ < 300000			
3					
4	Left-tailed Test				
5					
6	sample mean:	288000			
7	sample st. dev.:	80000			
8	sample size:	100			
9					
10	Significance level:	0.025			
11	Critical value:	-1.959961			
12					
13	Test statistic:	-1.5			
14					
15	p-value:	0.0668072			

Figure 9.5 Setting up and calculating the critical value (=NORMSINV(.025)), the test statistic (see formula bar above) and the p-value (=NORMSDIST(−1.5)) for this left-tailed hypothesis test. Fail to reject H₀. Not enough evidence to reject the claim.

9.3 HYPOTHESIS TESTS ABOUT A POPULATION MEAN: SMALL SAMPLES

In the case of a small sample (drawn from a population with a normal distribution and unknown standard deviation), the procedure is largely the same, except that we need to use the t-distribution instead of the normal distribution. In order to find the critical value(s) for an associated significance level, α, we will use Excel's TINV function, and to find the p-value, we will use the TDIST function.

Recall from Section 8.2 that the TINV function requires two inputs. The first is the value of α *for a confidence interval or a two-tailed test*. If we have a left- or right-tailed test, we need to enter α *multiplied by 2* in order to get the (positive) critical value. The second input is the degrees of freedom, which is one less than the sample size, $n–1$.

The TDIST function requires 3 inputs:
1) the test statistic, t, obtained from standardizing the sample statistics,
2) the degrees of freedom, $n–1$, and
3) whether the test is 1-tailed or 2-tailed (enter 1 or 2).

Example 9-6 Grand Auto Corporation produces auto batteries. The company claims that its top-of-the-line Never Die batteries are good, on average, for at least 65 months. A consumer protection agency tested 15 such batteries to check this claim. It found the mean life of these 15 batteries to be 63 months with a standard deviation of 2 months. At the 5% significance level, test the company's claim. (Assume that the life of such a battery has an approximately normal distribution so that the t-distribution is applicable.) Also calculate the p-value.

Solution: In order to set up the test, identify and enter the sample statistics into a blank Excel worksheet: $n = 15$, $\bar{x} = 63$, and $s = 2$. Note that the claim to be tested is $\mu \geq 65$.

Step 1. Enter the null and alternate hypotheses and note that the claim is the null hypothesis since it includes equality.

Step 2. Note that this is a left-tailed test, since the alternate hypothesis has the symbol "<" in it.

Step 3. Determine the critical value corresponding to the significance level of α = .05. Click on an empty cell and insert the TINV function. Since this is a 1-tailed test, we need to double α for the first input of the function (as if it were a two-tailed test or confidence interval). Enter 2*.05 (or .1). For the second input, the degrees of freedom, enter 15-1 (or 14). Since this is a left-tailed test, we need the negative of the resulting t-score of 1.7613. Double-click on the cell and insert a "-" after the "=" and just before the function name.

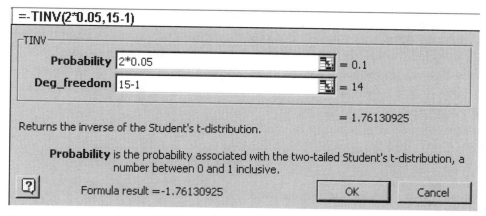

Figure 9.6 Using Excel's TINV function to obtain the critical value in a left-tailed test (small sample case).

Step 4. Calculate the value of the test statistic, i.e. standardize the sample statistics to find the associated t-score. Insert the STANDARDIZE function into a blank cell. Move the window out of the way, if necessary, and click on the cell with the sample mean of 63 in it for **X**, enter the hypothesized population mean of 65 for **Mean**, and enter 2/SQRT(15) for **Standard_dev** (or click on the cells that contain these values of μ, s and n instead of typing them in). You should obtain a value of –3.873.

Step 5. Since the test statistic of –3.873 is well beyond the critical value of –1.761 (in the rejection region), we must reject the null hypothesis (claim) and therefore conclude that there is enough evidence to conclude that the company's claim is false.

	B9 ▼	=	=STANDARDIZE(B2,B4,B3/SQRT(B1))			
	A	B	C	D	E	F
1	$n =$	15				
2	x-bar =	63				
3	$s =$	2				
4	$\mu \geq$	65	<- Null hypothesis = claim			
5	$\mu <$	65	<- Alternate hypothesis			
6						
7	critical value:	-1.7613				
8						
9	test statistic:	-3.873				
10						
11	Reject the null hypothesis!					

Figure 9.7 Calculating the test statistic via cell references to the values of x-bar, μ, s, and n in the STANDARDIZE function.

The p-value can then be found with the TDIST function. Click on a blank cell and enter this function. For the first input, enter the test statistic calculated earlier, but *you must enter its absolute value – this input must be positive.* For the second input, the degrees of freedom, enter 15-1 (or 14). And for the third input, enter a 1 since this is a one-tailed test. Click on **OK** or hit ENTER and you should see the *p*-value of 0.0008 appear. This is certainly smaller than α = .05! Looks like we would reject the null hypothesis (and thus the company's claim) even at the highly statistically significant level of α = .01.

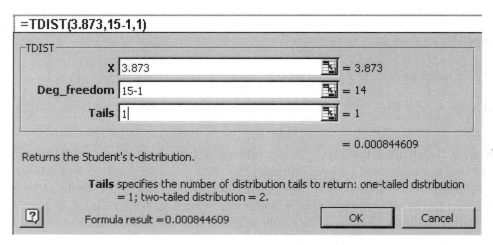

Figure 9.8 Using Excel's TDIST function to obtain the *p*-value for a one-tailed test (small sample case).

9.4 HYPOTHESIS TESTS ABOUT A POPULATION PROPORTION: LARGE SAMPLES

Hypothesis tests about a population proportion, *p*, can be done very similarly to those about a population mean, μ. In this section, though, we will standardize the sample statistic in order to get the test statistic by entering the formula

$$z = \frac{\hat{p} - p}{\sqrt{\dfrac{pq}{n}}}$$

into Excel, instead of using the STANDARDIZE function. Recall that $\hat{p} = \dfrac{x}{n}$ is the sample proportion (*x* is the number of elements in the sample with the certain characteristic). As long as the sample size is large, we can use the normal distribution and thus the functions NORMSINV and NORMSDIST in order to find the critical value(s) and *p*-value, respectively. (The criteria for "large sample" in this case are that *np* and *nq* are both greater than 5.)

Example 9-7 In a poll by the National Center for Women and Aging at Brandeis University, 51% of the women over 50 said that aging is not as bad as they had expected (*USA TODAY*, November 19, 2002). Assume that this result holds true for the 2002 population of all women aged 50 and over. In a recent random sample of 400 women aged 50 and over, 216 of them (54%) said that aging is not as bad as they had expected. Using the 1% significance level, can you conclude that the current percentage of women aged 50 and over who think that aging is not as bad as they had expected is different from that for 2002? Calculate the *p*-value.

Solution: In order to set up the test, identify and enter the sample statistics into a blank Excel worksheet: $n = 400$ and $\hat{p} = 216/400 = .54$. Note that the claim to be tested is that $p \neq .51$. Step 1. Enter the null and alternate hypotheses and note that the claim is the alternate hypothesis since it does not include equality. Since the hypothesized proportion is $p = .51$ (in the null hypothesis), we are also assuming that $q = 1-.51 = .49$. Type this value into another cell.

   Step 2. Note that this is a two-tailed test, since the alternate hypothesis has the symbol "$\neq$" in it.

   Step 3. Determine the negative critical value corresponding to the significance level of $\alpha = .01$ by typing "=NORMSINV(.01/2)" into a blank cell. (We must divide α by 2 since this is a two-tailed test.) Hit ENTER and you should see -2.57583 appear. Of course, $+2.57583$ is the other critical value.

   Step 4. Calculate the value of the test statistic, i.e. standardize the sample statistics to find the associated *z*-score. Here's where we will use the formula mentioned above. Click on an empty cell and type "=(" and click on the cell with the sample proportion, .54, in it. Then type "-" and click on the cell with the hypothesized proportion, .51, in it. Then type ")/SQRT(" and click on the cell with .51 in it again, type "*" and click on the cell with q, .49, in it, type "/" and click on the cell with n, 400, in it, type ")" and hit ENTER. You should see the value of 1.20024 appear. The formula bar should look like the one in Figure 9.9.

   Step 5. Since the test statistic of approximately 1.20 is not beyond the critical value of approximately 2.58 (it is not in the rejection region), we fail to reject the null hypothesis and therefore cannot accept the claim (the alternate hypothesis). We conclude that there is not enough evidence to support the claim that the current percentage of women aged 50 and over who think that aging is not as bad as they had expected is any different from what it was in 2002.

B11			=	=(B2-B4)/SQRT(B4*B7/B1)		
	A	B	C	D	E	
1	n =	400				
2	p-hat =	0.54				
3						
4	p =	0.51	<- Null hypothesis			
5	p <>	0.51	<- Alternate hypothesis (= claim)			
6						
7	q =	0.49				
8						
9	critical values:	-2.57583	and	2.575835		
10						
11	test statistic:	1.20024				

Figure 9.9 Using the test statistic formula for proportions in Excel, with cell references to the values of $\hat{p}$ and p, q, and n.

The p-value can be found by inserting the test statistic into the NORMSDIST function. However, since the test statistic is positive, we need to subtract the cumulative probability from 1 to get the area *to the right*. Then, since this is a two-tailed test, we need to double it. Click on a blank cell and type the following:

=2*(1-NORMSDIST(1.20)).

Hit ENTER, and you should see that the p-value is approximately 0.23, definitely not smaller than the significance level of $\alpha = .01 = 1\%$. In fact, we would not be able to reject the null hypothesis even with $\alpha = 20\%$! The difference in the percentages of women aged 50 and over who think that aging is not as bad as they had expected is just not statistically significant.

Exercises

9.1 A consumer advocacy group suspects that a local supermarket's 10-ounce packages of cheddar cheese actually weigh less than 10 ounces (on the average). The group took a random sample of 36 such packages and found that the mean weight for the sample was 9.955 ounces with a standard deviation of .15 ounce. Use Excel's NORMDIST function to find the p-value for the test of their claim. Using a significance level of $\alpha = 0.05$, what would be the conclusion? What would it be at $\alpha = 0.01$?

9.2 According to a survey by the National Retail Association, the average amount that households in the United States planned to spend on gifts, decorations, greeting cards, and food during the 2001 holiday season was $940 (*Money*, December 2001). Suppose that a recent random sample of 324 households showed that they plan to spend an average of $1005 on such items during this year's holiday season with a standard deviation of $330. Test at the 1% significance level whether the mean of such planned holiday-related expenditures for households for this year differs from $940. Use either NORMDIST or NORMSDIST (with the standardized test statistic) to calculate the *p*-value.

9.3 According to the Bureau of Labor Statistics, there were 8.1 million unemployed people aged 16 years and over in August 2002. The average duration of unemployment for these people was 16.3 weeks (*Bureau of Labor Statistics News,* September 6, 2002). Suppose that a recent random sample of 400 unemployed Americans aged 16 years and over gave a mean duration of unemployment of 16.9 weeks with a standard deviation of 4.2 weeks. Test at the 2% significance level whether the current mean duration of unemployment for all unemployed Americans aged 16 years and over exceeds 16.3 weeks. Either use NORMDIST or use NORMSDIST with the standardized test statistic in order to calculate the *p*-value.

9.4 According to a basketball coach, the mean height of all female college basketball players is 69.5 inches. A random sample of 25 such players produced a mean height of 70.25 inches with a standard deviation of 2.1 inches. Assuming that the heights of all female college basketball players are normally distributed, test at the 1% significance level whether their mean height is different from 69.5 inches. Use TDIST to calculate the *p*-value.

9.5 According to *Money* magazine, the average cost of a visit to a doctor's office in the United States was $60 in 2002 (*Money*, Fall 2002). Suppose that a recent random sample of 25 visits to doctors gave a mean of $63.50 and a standard deviation of $2.00. Using the 5% significance level, can you conclude that the current mean cost of a visit to a doctor's office exceeds $60? (Assume that such costs for all visits to doctors are normally distributed so that the *t*-distribution applies.) Use TDIST to calculate the *p*-value.

9.6 According to the Employment Policy Foundation's seventh annual report, titled *Challenges Facing the American Workplace,* women held 49% of management and professional jobs in 2000 (*The Hartford Courant,* September 2, 2002). Suppose that a recent random sample of 200 such jobs found that 104 of these (52%) are held by women. Can you conclude that the percentage of such jobs that are held by women currently exceeds 49%? Use Excel to calculate the critical value, test statistic, and *p*-value, and use $\alpha = .025$.

CHAPTER **10**

Estimation and Hypothesis Testing: Two Populations

10.1 INFERENCES ABOUT THE DIFFERENCES BETWEEN TWO POPULATION MEANS FOR LARGE AND INDEPENDENT SAMPLES

In this section, we will show how to use Excel to conduct a hypothesis test about the means of two separate populations, when we have independent and large samples from each population and the sample statistics have already been calculated. We will standardize the sample statistics in order to get the test statistic by entering into Excel the formula

$$z = \frac{\bar{x}_1 - \bar{x}_2}{\sqrt{\dfrac{\sigma_1^2}{n_1} + \dfrac{\sigma_2^2}{n_2}}}$$

where $\bar{x}_1$ and $\bar{x}_2$ are the sample means, σ_1 and σ_2 are the population standard deviations (which can be approximated by the sample standard deviations), and n_1 and n_2 are the sample sizes. Then we will use the functions NORMSINV and NORMSDIST in order to find the critical value(s) and p-value, respectively.

Example 10-1 According to the U.S. Bureau of the Census, the average annual salary of full-time state employees was $49,056 in New York and $46,800 in Massachusetts in 2001 (*The Hartford Courant*, December 5, 2002). Suppose that these mean salaries are based on random samples of 500 full-time state employees from New York and 400 full-time state employees from Massachusetts, and that the population standard deviations of the 2001 salaries of all full-time state employees in these two states were $9000 and $8500, respectively. Test at the 1% significance level that the 2001 mean salaries of full-time state employees in New York and Massachusetts are different. Calculate the p-value.

Solution: In order to set up the test, identify and organize the sample statistics into a blank Excel worksheet. (See Figure 10.1.) Note that the claim to be tested is that $\mu_1 \neq \mu_2$.

Step 1. Enter the null and alternate hypotheses and note that the claim is the alternate hypothesis since it does not include equality.

Step 2. Note that this is a two-tailed test, since the alternate hypothesis has the symbol "$\neq$" in it.

Step 3. Determine the negative critical value corresponding to the significance level of $\alpha = .01$ by typing: =NORMSINV(.01/2) into a blank cell. (We must divide α by 2 since this is a two-tailed test.) Hit ENTER and you should see -2.576 appear. Of course, $+2.576$ is the other critical value.

Step 4. Calculate the value of the test statistic, i.e. standardize the sample statistics to find the associated z-score. Here's where we will use the formula mentioned above. Click on an empty cell and type "=(" and click on the cell with the first sample's mean, 49056, in it. Then type "-" and click on the cell with the second sample's mean, 46800, in it. Then type ")/SQRT(" and click on the cell with the first standard deviation, 9000, in it, type "^2/" and click on the cell with the first sample's size, 500, in it, type "+" and click on the cell with the second standard deviation, 8500, in it, type "^2/" and click on the cell with the second sample's size, 400, in it, type ")" and hit ENTER. You should see the value of 3.8542 appear. The formula bar should look like the one in Figure 10.1.

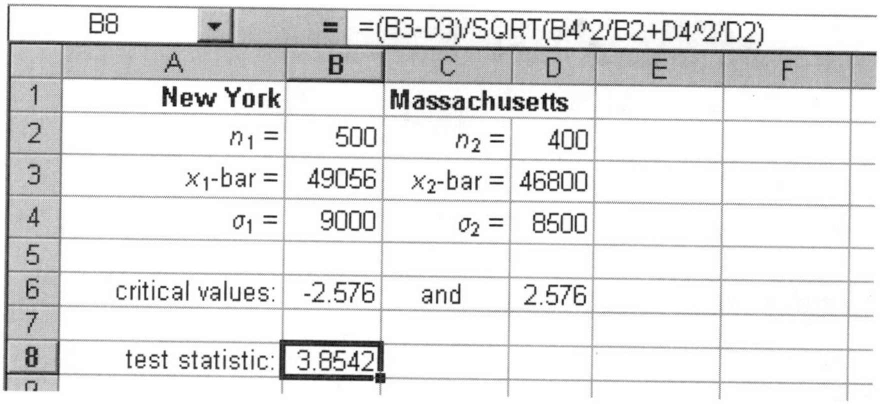

B8 = =(B3-D3)/SQRT(B4^2/B2+D4^2/D2)

	A	B	C	D	E	F
1	**New York**		**Massachusetts**			
2	n_1 =	500	n_2 =	400		
3	x_1-bar =	49056	x_2-bar =	46800		
4	σ_1 =	9000	σ_2 =	8500		
5						
6	critical values:	-2.576	and	2.576		
7						
8	test statistic:	3.8542				

Figure 10.1 Using the test statistic formula with references to the sample statistics organized in an Excel spreadsheet.

Step 5. Since the test statistic of approximately 3.85 is beyond the critical value of approximately 2.58 (it is in the rejection region), we reject the null hypothesis and therefore accept the claim (the alternate hypothesis). We conclude that the 2001 mean salaries of full-time state employees in New York and Massachusetts are different.

The *p*-value can be found as shown in Chapter 9, by inserting the test statistic into the NORMSDIST function. Since the test statistic is positive, we need to subtract the cumulative probability from 1 to get the area *to the right,* and since this is a two-tailed test, we then need to double it. Click on a blank cell and type the following:

=2*(1-NORMSDIST(3.85)).

Hit ENTER, and you should see that the *p*-value is approximately 0.0001, definitely smaller than the significance level of $\alpha = .01 = 1\%$. We have a highly statistically significant difference here!

10.2 INFERENCES ABOUT THE DIFFERENCES BETWEEN TWO POPULATION MEANS FOR SMALL AND INDEPENDENT SAMPLES

In this section, we will use and interpret the *t*-Test in Excel's Data Analysis ToolPak in order to test a hypothesis about the difference between two population means when *small* and independent samples are drawn from normally distributed populations. In order to use this tool, you must have the raw data, and not just the sample statistics.

After the raw data has been entered into an Excel spreadsheet, go to **Tools>Data Analysis** and simply choose the correct option for the situation: whether you are assuming the population variances (and thus the standard deviations) are equal or unequal.

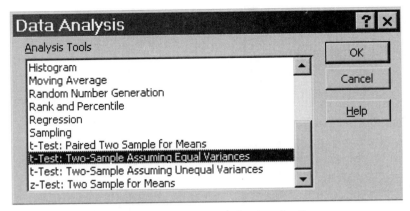

Figure 10.2 Select the appropriate t-Test for the situation.

Example 10-2 The following information was obtained from two independent samples selected from two normally distributed populations with unknown but equal standard deviations. Test at the 1% significance level if μ_1 is greater than μ_2.

Sample 1: 27   39   25   33   21   35   30   26   25   31   35   30   29

Sample 2: 24   28   23   25   24   22   29   26   29   28   19   29

Solution: Enter the data into an Excel spreadsheet, using one row (or column) for each sample. Go to **Tools>Data Analysis** and select **t-Test: Two-Sample Assuming Equal Variances**, as shown in Figure 10.2. (The variances will be equal if the standard deviations are equal, of course.) Recall that if **Data Analysis** is not an option on the **Tools** menu, then you need to go to **Tools>Add-Ins**, check the checkboxes next to **Analysis ToolPak** and **Analysis ToolPak –VBA,** and click on **OK.** Then go to **Tools** and you should see **Data Analysis.** (If this tool was not originally installed with Excel, then you will need to reinstall Excel in order to have it available.)

   In the **t-Test** window, highlight the first sample's data for the **Variable 1 Range** and the second sample's data for the **Variable 2 Range.** Then enter "0" for the **Hypothesized Mean Difference.** This is because the null hypothesis for this claim is $\mu_1 \leq \mu_2$ (the claim is that $\mu_1 > \mu_2$), and if they are equal, then their difference is 0. Finally, enter the level of significance where it says **Alpha.** Recall that we are to use a significance level of 1%, so enter 0.01.

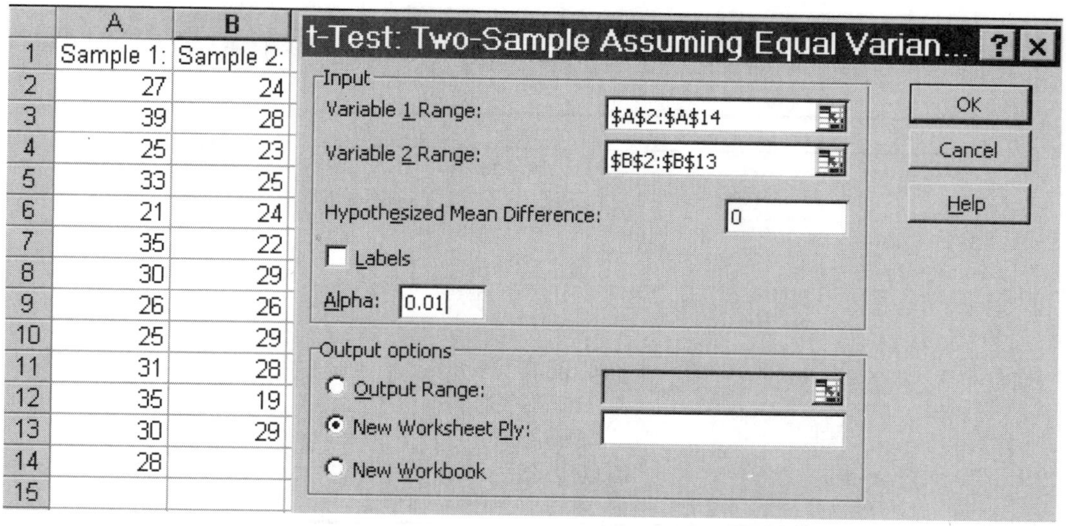

	A	B
1	Sample 1:	Sample 2:
2	27	24
3	39	28
4	25	23
5	33	25
6	21	24
7	35	22
8	30	29
9	26	26
10	25	29
11	31	28
12	35	19
13	30	29
14	28	
15		

t-Test: Two-Sample Assuming Equal Varian... ? X

Input
Variable 1 Range: $A$2:$A$14
Variable 2 Range: $B$2:$B$13
Hypothesized Mean Difference: 0
☐ Labels
Alpha: 0.01

Output options
○ Output Range:
◉ New Worksheet Ply:
○ New Workbook

OK
Cancel
Help

Figure 10.3 Enter the data ranges, the hypothesized mean of 0, and the significance level (α) for the *t*-test.

Click on **OK**, and a new worksheet will be created with several columns of information highlighted. Go to **Format>Column>AutoFit Selection** so that you can see it all. Now we need to pick out the important information and use it to draw the proper inference about the populations. Recall that the null hypothesis is $\mu_1 \leq \mu_2$ and that the claim is the alternate hypothesis, that $\mu_1 > \mu_2$. Notice that this inequality does hold for the sample means that are given on the first line that says **Mean**: $\bar{x}_1 = 29.615$ and $\bar{x}_2 = 25.5$. The question is whether this is statistically significant. Since this is a right-tailed test, look at the *p*-value for one tail, on the line that says **P(T<=t) one-tail**. It is approximately 0.0118, which is not less than 0.01 (though it is close). So we cannot reject the null hypothesis and therefore cannot support the claim that μ_1 is greater than μ_2.

Other pieces of information that you may find interesting from this display include the test statistic for the data that is shown on the line that says **t Stat**. It is approximately 2.42285. Also, the critical value for the significance level of 1% for a one-tailed test is shown on the line that says **t Critical one-tail**. It is approximately 2.49987. So you can see from this also that the test statistic is not in the rejection region.   Also notice the variances on the second line. Since they are quite different, it may be advantageous to repeat this test with the assumption of *unequal* variances.

	A	B	C
1	t-Test: Two-Sample Assuming Equal Variances		
2			
3		Variable 1	Variable 2
4	Mean	29.61538462	25.5
5	Variance	24.92307692	10.45454545
6	Observations	13	12
7	Pooled Variance	18.00334448	
8	Hypothesized Mean Difference	0	
9	df	23	
10	t Stat	2.422847775	
11	P(T<=t) one-tail	0.011839398	
12	t Critical one-tail	2.499873517	
13	P(T<=t) two-tail	0.023678795	
14	t Critical two-tail	2.807337296	

Figure 10.4 The output resulting from a *t*-test for two samples, assuming equal variances, using Excel's Data Analysis ToolPak.

10.3 INFERENCES ABOUT THE DIFFERENCES BETWEEN TWO POPULATION MEANS FOR PAIRED SAMPLES

Excel's Data Analysis ToolPak also has the option for a *t*-test to compare means in the case of paired (dependent) samples taken from an approximately normally distributed population.

Example 10-3 A researcher wanted to find the effect of a special diet on systolic blood pressure. She selected a sample of seven adults and put them on this dietary plan for three months. The following table gives the systolic blood pressures of these seven adults before and after the completion of this plan.

Before	210	180	195	220	231	199	224
After	193	186	186	223	220	183	233

Let μ_d be the mean of the differences between the systolic blood pressures before and after completing this special dietary plan for the population of all adults. Using the 5% significance level, can we conclude that the mean of the paired differences, μ_d, is different from zero? Assume that the population of paired differences is (approximately) normally distributed.

Solution: Enter the data into an Excel spreadsheet, using two rows (or columns), making sure that you maintain the pairings. Go to **Tools>Data Analysis** and select **t-Test: Paired Two Sample for Means**. (This is the one right above the selection shown in Figure 10.2.)

In the **t-Test** window, highlight the first sample's data for the **Variable 1 Range** and the second sample's data for the **Variable 2 Range**. Then enter "0" for the **Hypothesized Mean Difference**. Finally, check the level of significance where it says **Alpha**. The default is what we want in this case, 0.05 for the 5% level.

	A	B	t-Test: Paired Two Sample for Means	? x
1	Before	After		
2	210	193	Input	OK
3	180	186	Variable 1 Range: $A$2:$A$8	
4	195	186	Variable 2 Range: $B$2:$B$8	Cancel
5	220	223		
6	231	220	Hypothesized Mean Difference: 0	Help
7	199	183		
8	224	233	Labels	
9			Alpha: 0.05	
10				
11			Output options	
12			Output Range:	
13			New Worksheet Ply:	
14			New Workbook	
15				

Figure 10.5 Enter the data ranges, the hypothesized mean of 0, and the significance level (α) for the *t*-test.

Click on **OK**, and a new worksheet will be created with several columns of information highlighted. Go to **Format>Column>AutoFit Selection** so that you can see it all (or just widen column A).

The null hypothesis for this test is that $\mu_d = 0$, and the claim is the alternate hypothesis, that $\mu_d \neq 0$. So this is a two-tailed test. Look at the *p*-value for two tails, on the line that says **P(T<=t) two-tail**. It is approximately 0.265956, which is not less than 0.05. So we cannot reject the null hypothesis and therefore conclude that the mean of the differences between the systolic blood pressures before and after completing this special dietary plan for the population of all adults is not any different from (is equal to) zero.

Notice that the display also includes the test statistic for the data, shown on the line that says **t Stat**. It is approximately 1.226. Also, the critical value for the significance level of 5% for a two-tailed test is shown on the line that says **t Critical two-tail**. It is approximately 2.447. So you can see from this also that the test statistic is not in the rejection region.

	A	B	C
1	t-Test: Paired Two Sample for Means		
2			
3		Variable 1	Variable 2
4	Mean	208.4286	203.4286
5	Variance	327.619	444.2857
6	Observations	7	7
7	Pearson Correlation	0.85916	
8	Hypothesized Mean Difference	0	
9	df	6	
10	t Stat	1.226498	
11	P(T<=t) one-tail	0.132978	
12	t Critical one-tail	1.943181	
13	P(T<=t) two-tail	0.265956	
14	t Critical two-tail	2.446914	

Figure 10.6 The output resulting from a paired sample t-test in Excel's Data Analysis ToolPak.

10.4 INFERENCES ABOUT THE DIFFERENCES BETWEEN TWO POPULATION PROPORTIONS FOR LARGE AND INDEPENDENT SAMPLES

In this section, we will go back to the approach we used in Section 10.1 in order to conduct a hypothesis test about the difference between two population proportions, when we have independent and large samples from each populations and the sample statistics have already been calculated. We will standardize the sample statistics in order to get the test statistic by entering into Excel the formula

$$z = \frac{\hat{p}_1 - \hat{p}_2}{\sqrt{\overline{p}\overline{q}\left(\frac{1}{n_1} + \frac{1}{n_2}\right)}}$$

where $\overline{p} = \frac{x_1 + x_2}{n_1 + n_2}$ (or $\overline{p} = \frac{n_1\hat{p}_1 + n_2\hat{p}_2}{n_1 + n_2}$ if you don't know x_1 and x_2) and $\overline{q} = 1 - \overline{p}$. Recall that $\hat{p}_1 = \frac{x_1}{n_1}$ and $\hat{p}_2 = \frac{x_2}{n_2}$ are the sample proportions, n_1 and n_2 are the sample sizes, and x_1 and x_2 are the *number* of elements in each of the two samples that possess the certain characteristic. Then we can use the functions NORMSINV and NORMSDIST in order to find the critical value(s) and p-value, respectively.

Example 10-4 A researcher wanted to estimate the difference between the percentages of users of two toothpastes who will never switch to another toothpaste. In a sample of 500 users of Toothpaste A taken by this researcher, 100 said that they will never switch to another toothpaste. In another sample of 400 users of Toothpaste B taken by the same researcher, 68 said that they will never switch to another toothpaste. At the 1% significance level, can we conclude that the proportion of users of Toothpaste A who will never switch to another toothpaste is higher than the proportion of users of Toothpaste B who will never switch to another toothpaste?

Solution: In order to set up the test, identify and organize the sample statistics n_1 and x_1, and n_2 and x_2 into a blank Excel worksheet. Then have Excel calculate the corresponding sample proportions as shown in Figure 10.7. Note that the claim to be tested is: $p_1 > p_2$.

B4		$\blacktriangledown$	$=$	=B3/B2	
	A	B	C	D	E
1	Toothpaste A		Toothpaste B		
2	$n_1 =$	500	$n_2 =$	400	
3	$x_1 =$	100	$x_2 =$	68	
4	p_1-hat $=$	0.2	p_2-hat $=$	0.17	

Figure 10.7 Enter the sample statistics into Excel and have it calculate the sample proportions. (For subscripts, highlight the numeral to be subscripted and go to **Format>Cells** and click on the checkbox next to **Subscript**.)

Step 1. Enter the null and alternate hypotheses and note that the claim is the alternate hypothesis since it does not include equality.

Step 2. Note that this is a right-tailed test, since the alternate hypothesis has the symbol ">" in it.

Step 3. Determine the negative critical value corresponding to the significance level of $\alpha = .01$ by typing: =NORMSINV(.01) into a blank cell. Hit ENTER and you should see −2.32634 appear. Since this is a right-tailed test, the critical value we want is actually +2.32634.

Step 4. Calculate the value of the test statistic, i.e. standardize the sample statistics to find the associated z-score. Here's where we will use the formula mentioned above. First, calculate the "pooled sample proportion," $\overline{p}$. Click on an empty cell and type "=(" and click on the cell with the first sample's x-value, 100, in it. Then type "+" and click on the cell with the second sample's x-value, 68, in it. Then type ")/(" and click on the cell with the first sample's size, 500, in it, type "+" and click on the cell with the second sample's size, 400, in it, type ")" and hit ENTER. You should see the value of 0.186667 appear. With statistics entered as in Figure 10.7 above, the formula bar should read: =(B3+D3)/(B2+D2).

Now for $\bar{q}$, click on the cell right below this value, type "=1-" and arrow up to the cell above it (with $\bar{p}$ in it) and hit ENTER. You should see the value of 0.813333 appear. Finally, click on the cell below that for the test statistic. Type "=(" and click on the cell with the first sample's proportion, 0.2, in it. Then type "-" and click on the cell with the second sample's proportion, 0.17, in it. Then type ")/SQRT(" and click on the cell with $\bar{p}$, 0.186667, in it, type "*" and click on the cell with $\bar{q}$, 0.813333, in it, type "*(1/" and click on the cell with the first sample's size, 500, in it, type "+1/" and click on the cell with the second sample's size, 400, in it, type "))" and hit ENTER. You should see the value of 1.14775 appear. The formula bar should look like the one in the figure below.

B10	▼	=	=(B4-D4)/SQRT(B8*B9*(1/B2+1/D2))		
	A	B	C	D	E	F
1	Toothpaste A		Toothpaste B			
2	$n_1 =$	500	$n_2 =$	400		
3	$x_1 =$	100	$x_2 =$	68		
4	p_1-hat =	0.2	p_2-hat =	0.17		
5						
6	critical value:	-2.32634	--->	2.326342	for a right-tailed test	
7						
8	p-bar =	0.186667				
9	q-bar =	0.813333				
10	test statistic:	1.14775				
11						

Figure 10.8 Using the test statistic formula with references to the sample statistics and other calculations organized in an Excel spreadsheet.

Step 5. Since the test statistic of approximately 1.15 is not beyond the critical value of approximately 2.33 (it is not in the rejection region), we fail to reject the null hypothesis. Therefore, we conclude that the proportion of users of Toothpaste A who will never switch to another toothpaste is probably not any greater than the proportion of users of Toothpaste B who will never switch to another toothpaste.

The p-value can be found as before by inserting the test statistic into the NORMSDIST function. Since the test statistic is positive, we need to subtract the cumulative probability from 1 to get the area *to the right*. Click on a blank cell and type the following:

=1-NORMSDIST(1.15).

Hit ENTER, and you should see that the p-value is approximately 0.1251, definitely larger than the significance level of $\alpha = .01 = 1\%$. The results are not even statistically significant at the 10% level!

Exercises

10.1   A business consultant wanted to investigate if providing day-care facilities on premises by companies reduces the absentee rate of working mothers with six-year-old or younger children. She took a sample of 45 such mothers from companies that provide day-care facilities on premises. These mothers missed an average of 6.4 days from work last year with a standard deviation of 1.20 days. Another sample of 50 such mothers taken from companies that do not provide day-care facilities on premises showed that these mothers missed an average of 9.3 days last year with a standard deviation of 1.85 days. Using the 2.5% significance level, can you conclude that the mean number of days missed per year by mothers working for companies that provide day-care facilities on premises is less than the mean number of days missed per year by mothers working for companies that do not provide day-care facilities on premises? Use Excel to calculate the critical value, test statistic, and p-value.

10.2   The following information was obtained from two independent samples selected from two normally distributed populations with unknown but equal standard deviations. Use Excel's Data Analysis ToolPak in order to test at the 2.5% significance level if μ_1 is lower than μ_2.

Sample 1:	13	14	9	12	8	10	5	10	9	12	16	
Sample 2:	16	18	11	19	14	17	13	16	17	18	22	12

10.3   Several retired bicycle racers are coaching a large group of young prospects. They randomly select seven of their riders to take part in a test of the effectiveness of a new dietary supplement that is supposed to increase strength and stamina. Each of the seven riders does a time trial on the same course. Then they all take the dietary supplement for four weeks. All other aspects of their training program remain as they were prior to the time trial. At the end of the four weeks, these riders do another time trial on the same course. The times (in minutes) recorded by each rider for these trials, before and after the four-week period, are shown in the following table.

Before	103	97	111	95	102	96	108
After	100	95	104	101	96	91	101

Use Excel's Data Analysis ToolPak in order to test at the 2.5% significance level whether taking this dietary supplement results in faster times in the time trials. Assume that the population of paired differences is (approximately) normally distributed.

10.4   A company claims that its 12-week special exercise program significantly reduces weight. A random sample of six persons was selected, and these persons were put on this exercise program for 12 weeks. The following table gives the weights (in pounds) of those six persons before and after the program.

Before	180	195	177	221	208	199
After	183	187	161	204	197	189

Use Excel's Data Analysis ToolPak in order to test at the 1% significance level whether the mean weight loss for all persons due to this special exercise program is greater than zero. Assume that the population of paired differences is (approximately) normally distributed.

10.5   The lottery commissioner's office in a state wanted to find if the percentages of men and women who play the lottery often are different. A sample of 500 men taken by the commissioner's office showed that 160 of them play the lottery often. Another sample of 300 women showed that 66 of them play the lottery often. Testing at the 1% significance level, can you conclude that the proportions of all men and all women who play the lottery often are different? Use Excel to calculate the critical value, test statistic, and p-value.

CHAPTER **11**

Chi-Square Tests

CHAPTER OUTLINE
11.1 A Goodness-of-Fit Test
11.2 A Test of Independence or Homogeneity

11.3 Inferences about the Population Variance

11.1 A GOODNESS-OF-FIT TEST

In this section, we will use the Chi-Square distribution built into Excel's CHITEST and CHIINV functions in order to test how "good" the observed frequencies for a multinomial experiment "fit" an expected pattern of frequencies.

The CHITEST function has two *array* inputs:
1) the array (e.g. column or row) of observed frequencies, and
2) the corresponding array of expected frequencies.

Its output is the *p*-value of the hypothesis test, which you can then compare to the significance level, α, that is desired. (If it's less than α, you reject the null hypothesis.)

The CHIINV function has two single-valued inputs:
1) the area of the right tail in a chi-square distribution for which you want the value of χ^2, and
2) the degrees of freedom, which is one less than the number of categories, $k - 1$.

This function will give you the critical value if you enter α for the first input, and it will give you the test statistic if you enter the *p*-value for the first input. (A goodness-of-fit test is always a right-tailed test.)

Example 11-1 A bank has an ATM installed inside the bank, and it is available to its customers only from 7 AM to 6 PM Monday through Friday. The manager of the bank wanted to investigate if the percentage of people who use this ATM is the same for each of the five days (Monday through Friday) of the week. She randomly selected one week and counted the number of people who used this ATM on each of the five days during this week. The information she obtained is given in the following table, where the number of users represents the number of transactions on this ATM on these days. For convenience, we will refer to these transactions as "people" or "users."

Day	Monday	Tuesday	Wednesday	Thursday	Friday
Number of users	253	197	204	279	267

At the 1% level of significance, can we reject the null hypothesis that the proportion of people who use this ATM each of the five days of the week is the same? Assume that this week is typical of all weeks in regard to the use of this ATM.

Solution: First, enter the table of observed frequencies into two rows of an Excel spreadsheet. Below each observed frequency, enter the expected frequency. In this case, the expected frequencies would all be the same: the total of the frequencies divided by 5 (or the total times 0.2). Have Excel calculate the sum of the frequencies with the SUM function. (Click on an empty cell, type "=SUM(" and highlight the cells containing the frequencies and then hit ENTER.) You should see that the total number of users in this sample is 1200. So in the cell below Monday's observed frequency, type "=1200/5" or "=1200*.2." Hit ENTER and you should see 240 appear. Copy this across to the cells below the observed frequencies of the other days of the week. (Drag the square in the lower right-hand corner to the right.)

Now we will use the CHITEST function on the arrays of the observed and expected frequencies. Click on an empty cell, click on the f_x icon, and insert CHITEST. Move the window out of the way, if necessary, and highlight the expected frequencies for the **Actual range** text box. Then click into (or TAB to) the **Expected range** text box and highlight the expected frequencies. (See Figure 11.1) Hit ENTER or click on **OK** and you should see the p-value of 0.000116 appear. Since this is less than $\alpha = .01$, we reject the null hypothesis and conclude that the proportion of persons who use this ATM is not the same for the five days of the week. A higher percentage of users of this ATM use this machine on one or more of these days.

If you want to calculate the test statistic and critical value, use Excel's CHIINV function. In this situation, there are 4 (= 5 categories – 1) degrees of freedom. So type "=CHIINV(.000116, 4)" and hit ENTER in order to get the test statistic, χ^2, of 25.18579. Type "=CHIINV(.01, 4)" and hit ENTER in order to get the critical value of 13.2767 for the significance level of $\alpha = .01$. Here, also, you can see that the test statistic is well into the right-tailed rejection region.

CHITEST	▼	X ✓ =	=CHITEST(B2:F2,B3:F3)		

	A	B	C	D	E	F
1	Day	Monday	Tuesday	Wednesday	Thursday	Friday
2	Number of users	253	197	204	279	267
3	Expected Frequencies	240	240	240	240	240
4						

CHITEST

Actual_range B2:F2| 📭 = {253,197,204,279,2

Expected_range B3:F3   📭 = {240,240,240,240,2

= 0.000116382

Returns the test for independence: the value from the chi-squared distribution for the statistic and the appropriate degrees of freedom.

    Actual_range is the range of data that contains observations to test against expected values.

[?]   Formula result =0.000116382   [OK]   [Cancel]

Figure 11.1 Using Excel's CHITEST function in order to find the *p*-value for a goodness-of-fit test.

B9	▼	=	=CHIINV(0.01,4)			

	A	B	C	D	E	F	G
1	Day	Monday	Tuesday	Wednesday	Thursday	Friday	Total
2	Number of users	253	197	204	279	267	1200
3	Expected Frequencies	240	240	240	240	240	
4							
5	Chi-test:	0.000116					
6							
7	Test statistic:	23.18579					
8							
9	Critical value for 1% level:	13.2767					
10							

Figure 11.2 Using Excel's CHIINV function in order to find the critical value for a goodness-of-fit test.

11.2 A TEST OF INDEPENDENCE OR HOMOGENEITY

A test of independence or homogeneity requires comparing the observed frequencies in a contingency table with expected frequencies that are calculated by multiplying the corresponding row and column totals and dividing by the sample size. Again, we will use CHITEST in order to obtain the *p*-value for either of these types of tests, and we can use CHIINV if we also want the test statistic and critical value. The degrees of freedom are found by taking one less than the number of rows times one less than the number of columns in the contingency table (not counting labels or totals). Like goodness-of-fit tests, tests of independence and homogeneity are always right-tailed tests. An example of a test of independence is shown below; the procedure for a test of homogeneity is completely analogous.

Example 11-2 Violence and lack of discipline have become major problems in schools in the United States. A random sample of 300 adults was selected, and they were asked if they favor giving more freedom to schoolteachers to punish students for violence and lack of discipline. The two-way classification of the responses of these adults is presented in the following table.

	In Favor	Against	No Opinion
Men	93	70	12
Women	87	32	6

Does the sample provide sufficient information to conclude that the two attributes, gender and opinions of adults, are dependent? Use a 1% significance level.

Solution: Enter the contingency table into a new Excel spreadsheet. Have Excel calculate the row and column totals by clicking on the cell below each column (and to the right of each row), typing "=SUM(" and highlighting the column (or row) and hitting ENTER. (Actually, you can just do the first row's sum and copy it down, and do the first column's sum and copy it across!)

B4	▼	= =SUM(B2:B3)		
	A	B	C	D	E
1		In Favor	Against	No Opinion	Totals
2	Men	93	70	12	175
3	Women	87	32	6	125
4	Totals	180	102	18	300

Figure 11.3 Use Excel's SUM function to calculate row and column totals for a contingency table.

Now click on an empty cell below the contingency table in order to calculate the expected frequency for the cell in the upper left-hand corner (Men In Favor). Type "=" and click on the cell with the first row's total, 175, in it, type "*" and click on the cell with the first column's total, 180, in it, type "/" and click on the cell with the total sample size, 300, in it, and hit ENTER. You should see the expected frequency of 105 appear.

In order to get the rest of the expected frequencies, we can copy this first one to the rest of the 3 columns and 2 rows, but we *first* have to make the cell references *partially absolute*. We want the columns to change for the column totals, but not the rows, and we want the rows to change for the row totals, but not the columns. So double-click on the cell with the first expected frequency in it (that was just calculated) in order to edit it, and insert $ signs before the column-reference (the letter) for the row total and before the row-reference (the number) for the column total, and before *both* references for the sample size (which we want to entirely absolute). See Figure 11.4. Now you can copy this cell to locations corresponding to the observed frequencies in the contingency table in order to get all of the corresponding expected frequencies!

	B7		▼	**=**	=$E2*B$4/$E$4	
	A	B	C	D	E	
1		In Favor	Against	No Opinion	Totals	
2	Men	93	70	12	175	
3	Women	87	32	6	125	
4	Totals	180	102	18	300	
5						
6	Expected Frequencies:					
7		105	59.5	10.5		
8		75	42.5	7.5		
9						

Figure 11.4 Make references partially absolute by inserting $ signs so that you can copy the formula for the expected frequencies and have only the row or column reference change.

Now we are ready for the CHITEST function. Click on an empty cell, click on the f_x icon, and insert CHITEST. Move the window out of the way, if necessary, and highlight the two rows and three columns of observed frequencies for the **Actual range** text box. *Make sure that you do NOT highlight the totals.* Then click into (or TAB to) the **Expected range** text box and highlight the expected frequencies. (See Figure 11.5.) Hit ENTER or click on **OK** and you should see the *p*-value of 0.016141 appear. Since this is not less than α = .01, we fail to reject the null hypothesis and state that there is not enough evidence from the sample to conclude that the two characteristics, *gender* and *opinions of adults*, are dependent for this issue.

     As in the previous example, if you want to calculate the test statistic and critical value, use Excel's CHIINV function. The degrees of freedom are (2-1)(3-1) = 2, since there are 2 rows and 3 columns of observed frequencies in the contingency table. So type "=CHIINV(.016141, 2)" and hit ENTER in order to get the test statistic, χ^2, of approximately 8.2527. Type "=CHIINV(.01, 2)" and hit ENTER in order to get the critical value of 9.210 for the significance level of α = .01. (Or click on the cell containing the *p*-value or α for the first input and enter the expression "(2-1)*(3-1)" for the second input as shown in the formula bar in Figure 11.6.) Here, also, you can see that the test statistic is not in the right-tailed rejection region.

CHITEST	▾	✕ ✓	=	=CHITEST(B2:D3,B7:D8)			
	A	B	C	D	E	F	G
1		In Favor	Against	No Opinion	Totals		
2	Men	93	70	12	175		
3	Women	87	32	6	125		
4	Totals	180	102	18	300		
5							
6	Expected Frequencies:						
7		105	59.5	10.5	.		
8		75	42.5	7.5			
9							

```
┌CHITEST──────────────────────────────────────────────────────┐
│                                                              │
│  Actual_range  [B2:D3|]                      ▥ = {93,70,12;87,32,6}
│                                                              │
│  Expected_range  [B7:D8]                     ▥ = {105,59.5,10.5;75,ᵉ
│                                                              │
│                                                  = 0.016141099
│  Returns the test for independence: the value from the chi-squared distribution for the
│  statistic and the appropriate degrees of freedom.
│      Actual_range is the range of data that contains observations to test against expected
│                   values.
│                                                              │
│  [?]      Formula result =0.016141099        [   OK   ]  [  Cancel  ]
└──────────────────────────────────────────────────────────────┘
```

Figure 11.5 Using Excel's CHITEST function in order to find the *p*-value for a test of independence.

A12	▾	=	=CHIINV(A10,(2-1)*(3-1))	
	A	B	C	D	E
10	0.016141	= p-value			
11					
12	8.252747	= test statistic			
13					
14	9.210351	= critical value			

Figure 11.6 Using Excel's CHIINV function in order to find the test statistic for a test of independence.

(*Note:* If the calculated *p*-value is extremely small, then it will be given in scientific notation, and Excel may not be able to retrieve the corresponding test statistic.)

11.3 INFERENCES ABOUT THE POPULATION VARIANCE

We will use the chi-square distribution for a test about a population variance as well. In this case, we will first calculate the test statistic using the formula

$$\chi^2 = \frac{(n-1)s^2}{\sigma^2}$$ (one less than the sample size times the sample variance divided by

the population variance). Then we will use the CHIINV function in order to calculate the test statistic and the critical value for a given level of significance. The degrees of freedom are one less than the sample size, $n-1$.

The 5-step procedure will be reminiscent of those used in previous chapters.

Example 11-3 One type of cookie manufactured by Haddad Food Company is Cocoa Cookies. The machine that fills packages of these cookies is set up in such a way that the average net weight of these packages is 32 ounces with a variance of .015 square ounce. From time to time the quality control inspector at the company selects a sample of a few such packages, calculates the variance of the net weights of these packages, and makes a test of hypothesis about the population variance. She always uses $\alpha = .01$. The acceptable value of the population variance is .015 square ounce or less. If the conclusion from the test of hypothesis is that the population variance is not within the acceptable limit, the machine is stopped and adjusted. A recently taken random sample of 25 packages from the production line gave a sample variance of .029 square ounce. Based on this sample information, do you think the machine needs an adjustment? (Assume that the net weights of cookies in all packages are normally distributed so that the χ^2-distribution applies.)

Solution: In order to set up the test, identify and enter the sample statistics into a blank Excel worksheet: $n = 25$ and $s^2 = .029$.

Step 1. Enter the null and alternate hypotheses:
  $\sigma^2 \leq .015$ (acceptable), and
  $\sigma^2 > .015$ (machine needs to be adjusted). Type these in.

Step 2. Note that this is a right-tailed test, since the alternate hypothesis has the symbol ">" in it.

Step 3. Determine the critical value corresponding to the significance level of $\alpha = .01$ by typing "=CHIINV(.01, 24)" into a blank cell. (Recall that the degrees of freedom are one less than the sample size of 25.) Hit ENTER and you should see 42.97978 appear.

Step 4. Calculate the value of the test statistic. Here's where we will use the formula mentioned above. Click on an empty cell and type "=24*.029/.015" and hit ENTER. (Or you can click on the cells containing the relevant values instead of typing them in if you wish.) You should see the value of 46.4 appear.

Step 5. Since the test statistic of 46.4 is beyond the critical value of approximately 42.98 (it is in the rejection region), we reject the null hypothesis and therefore conclude that the population variance is not within the acceptable limit. The machine should be stopped and adjusted.

B8			=	=24*0.029/0.015	
	A	B	C	D	E
1	$n =$	25		$s^2 =$	0.029
2					
3	$\sigma^2 \leq .015$	Null hypothesis			
4	$\sigma^2 > .015$	Alternate hypothesis			
5			CHIINV(.01,24)		
6	critical value:	42.97978			
7					
8	test statistic:	46.4			

Figure 11.7 Using the test statistic formula (shown in the formula bar above) and the CHIINV function (shown in the comment above) in order to test a hypothesis about a population variance. (In order to insert a comment, simply go to **Insert>Comment**.) The null hypothesis is rejected.

Exercises

11.1   The following table lists the frequency distribution of cars sold at an auto dealership during the past 12 months.

Month	Jan	Feb	Mar	Apr	May	Jun	Jul	Aug	Sep	Oct	Nov	Dec
Cars sold	23	17	15	10	14	12	13	15	23	26	27	29

Using the 10% significance level, will you reject the null hypothesis that the number of cars sold at this dealership is the same for each month? Use Excel's CHITEST and CHIINV functions to calculate the p-value, the test statistic, and the critical value.

11.2   One hundred auto drivers who were stopped by police for some violation were also checked to see if they were wearing their seat belts. The following table records the results of this survey.

	Wearing Seat Belt	Not Wearing Seat Belt
Men	34	21
Women	32	13

Test at the 2.5% significance level whether being a man or a woman and wearing or not wearing a seat belt are related. Use Excel's CHITEST and CHIINV functions to calculate the p-value, the test statistic, and the critical value.

11.3 National Electronics Company buys parts from two subsidiaries. The quality control department at this company wanted to check if the distribution of good and defective parts is the same for the supplies of parts received from both subsidiaries. The quality control inspector selected a sample of 300 parts received from Subsidiary A and a sample of 400 parts received from Subsidiary B. These parts were checked for being good or defective. The following table records the results of this investigation.

	Subsidiary A	Subsidiary B
Good	284	381
Defective	16	19

Using the 5% significance level, test the null hypothesis that the distributions of good and defective parts are the same for both subsidiaries. Use Excel's CHITEST and CHIINV functions to calculate the p-value, the test statistic, and the critical value.

11.4 An auto manufacturing company wants to estimate the variance of miles per gallon for its auto model AST727. A random sample of 22 cars of this model showed that the variance of miles per gallon for these cars is .62. Test at the 1% significance level whether the sample result indicates that the population variance is different from .30. Use Excel's CHIINV function to calculate the critical value and test statistic.

CHAPTER **12**

Analysis of Variance

CHAPTER OUTLINE

12.1 THE *F* DISTRIBUTION

Excel has the *F* distribution built in to its FDIST and FINV functions.

FDIST requires three inputs:
1) the *F* value for which you want the area of the right tail beyond it,
2) the degrees of freedom for the numerator, and
3) the degrees of freedom for the denominator.

It can be used to return the *p*-value for a given test statistic.

FINV also requires three inputs:
1) the probability or area of the right tail beyond the *F* value that you want,
2) the degrees of freedom for the numerator, and
3) the degrees of freedom for the denominator.

It can be used to return the critical value for a given level of significance.

Example 12-1 Find the *F* value for 8 degrees of freedom for the numerator, 14 degrees of freedom for the denominator (i.e. $df = (8, 14)$), and .05 area in the right tail of the *F* distribution curve.

Solution: Click on an empty cell in an Excel worksheet, click on the f_x icon, and insert FINV. Enter the three inputs .05, 8, and 14 and click on **OK**. (Or you can simply type "=FINV(.05, 8, 14)" and hit ENTER.) You should see the *F* value of 2.69867 appear.

Notice that you could go back the other way, if you wish, from the F-value back to the area. If you type "=FDIST(2.69867, 8, 14)" and hit ENTER, you'll get .05!

12.2 ONE-WAY ANALYSIS OF VARIANCE

In this section, we will use and interpret the one-way ANOVA tool in Excel's Data Analysis ToolPak in order to conduct a hypothesis test involving the means of several populations.

Example 12-2 Fifteen fourth-grade students were randomly assigned to three groups to experiment with three different methods of teaching arithmetic. At the end of the semester, the same test was given to all 15 students. The table gives the scores of students in the three groups.

Method I	Method II	Method III
48	55	84
73	85	68
51	70	95
65	69	74
87	90	67

At the 1% significance level, can we reject the null hypothesis that the mean arithmetic score of all fourth-grade students taught by each of these three methods is the same? (Assume that all the assumptions required to apply the one-way ANOVA procedure hold true.)

Solution: Enter the table above into a new Excel spreadsheet, including the column labels. Go to **Tools>Data Analysis** and select the very first option on the list: **Anova: Single Factor.**

Recall that if **Data Analysis** is not an option on the **Tools** menu, then you need to go to **Tools>Add-Ins**, check the checkboxes next to **Analysis ToolPak** and **Analysis ToolPak –VBA,** and click on **OK.** Then go to **Tools** and you should see **Data Analysis.** (If this tool was not originally installed with Excel, then you will need to reinstall Excel in order to have it available.)

Move the window out of the way, if necessary, and highlight the entire table of values for the **Input Range.** If you have entered the table as it is displayed above (and shown in Figure 12.1), then make sure that the radio button next to **Columns** is marked, for how the data is grouped. Since we are also highlighting the column headings in the first row, check the box next to **Labels in First Row.** This will make the output more readable. Finally, enter the significance level of .01 into the box next to where it says **Alpha**.

Click on **OK**, and a new worksheet will be created with several columns of information highlighted. Go to **Format>Column>AutoFit Selection** so that you can see it all. (Or just widen column A by dragging its right border further to the right.)

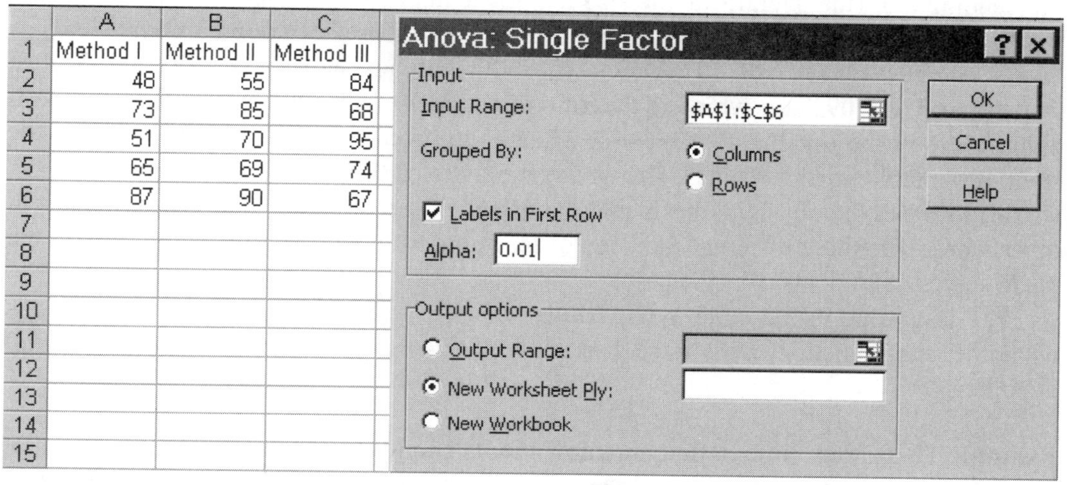

	A	B	C
1	Method I	Method II	Method III
2	48	55	84
3	73	85	68
4	51	70	95
5	65	69	74
6	87	90	67

Anova: Single Factor

Input

Input Range: `$A$1:$C$6`

Grouped By: ○ Columns ○ Rows

☑ Labels in First Row

Alpha: `0.01`

Output options

○ Output Range:

● New Worksheet Ply:

○ New Workbook

OK   Cancel   Help

Figure 12.1 Using the one-way ANOVA tool in Excel's Data Analysis ToolPak.

	A	B	C	D	E	F	G
1	Anova: Single Factor						
2							
3	SUMMARY						
4	Groups	Count	Sum	Average	Variance		
5	Method I	5	324	64.8	258.2		
6	Method II	5	369	73.8	194.7		
7	Method III	5	388	77.6	140.3		
8							
9							
10	ANOVA						
11	Source of Variation	SS	df	MS	F	P-value	F crit
12	Between Groups	432.1333	2	216.0667	1.092717	0.366464	6.926598
13	Within Groups	2372.8	12	197.7333			
14							
15	Total	2804.933	14				

Figure 12.2 The output resulting from the one-way ANOVA tool in Excel's Data Analysis ToolPak includes an ANOVA table.

The first table simply gives some summary statistics of each group separately. For example, you can look at the averages (means) of the groups in order to see how different they appear.   (The question, of course, is whether the difference is statistically significant.) The second table is the ANOVA table. The second column in this table gives the *between-samples sum of squares* (SSB) of 432.1333 and the *within-samples sum of squares* (SSW) of 2372.8. The third column gives the degrees of freedom for the numerator (one less than the number of groups, $3 - 1 = 2$) and the degrees of freedom for the denominator (the overall sample size minus the number of groups, $15 - 3 = 12$). The fourth column gives the *variance between samples* (MSB) of 216.0667 and the *variance within samples* (MSW) of 197.7333. The fifth column shows the value of the test statistic of $F = 1.092717$, which is the ratio of the MSB over the MSW. Then the sixth column shows the *p*-value of 0.366464, which is the area under the F distribution to the right of the test statistic. This is what you compare to α. Since it is much larger than .01, we fail to reject the null hypothesis that the means of the three populations are equal. In other words, the three different methods of teaching arithmetic do not seem to affect the mean scores of students.

The last value in the table is the critical value, $F = 6.926598$, for $\alpha = .01$. This is what the test statistic would have had to be *larger* than in order to reject the null hypothesis.

Example 12-3 From time to time, unknown to its employees, the research department at Post Bank observes various employees for their work productivity. Recently this department wanted to check whether the four tellers at a branch of this bank serve, on average, the same number of customers per hour. The research manager observed each of the four tellers for a certain number of hours. The following table gives the number of customers served by the four tellers during each of the observed hours.

Teller A	Teller B	Teller C	Teller D
19	14	11	24
21	16	14	19
26	14	21	21
24	13	13	26
18	17	16	20
	13	18	

At the 5% significance level, test the null hypothesis that the mean number of customers served per hour by each of these four tellers is the same. Assume that all the assumptions required to apply the one-way ANOVA procedure hold true.

Solution: Enter the table above into a new Excel spreadsheet, including the column labels. Go to **Tools>Data Analysis** and select **Anova: Single Factor.**

Move the window out of the way, if necessary, and highlight the entire table of values for the **Input Range**. It's okay to include the blank cells in the lower corners – they will be ignored. If you have entered the table as it is displayed above (and shown in Figure 12.3), then make sure that the radio button next to **Columns** is marked, for how the data is grouped. Since we are also highlighting the column headings in the first row, check the box next to **Labels in First Row**. Finally, check the significance level where it says **Alpha**. This needs to be .05 (which is the default) since we're using the 5% level for this test.

Click on **OK**, and a new worksheet will be created with several columns of information highlighted. Go to **Format>Column>AutoFit Selection** so that you can see it all. (Or just widen column A.)

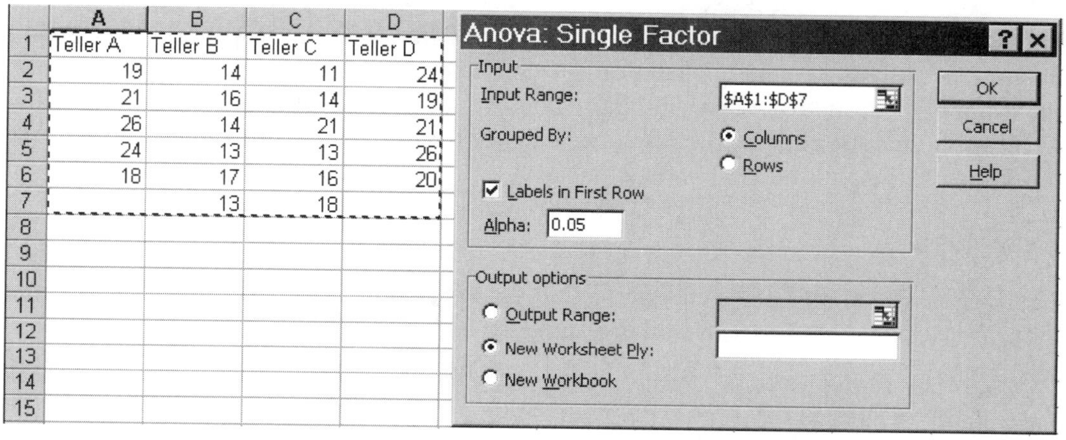

Figure 12.3 Using the one-way ANOVA tool in Excel's Data Analysis ToolPak. Don't worry about including empty cells when the groups are different sizes.

	A	B	C	D	E	F	G
1	Anova: Single Factor						
2							
3	SUMMARY						
4	*Groups*	*Count*	*Sum*	*Average*	*Variance*		
5	Teller A	5	108	21.6	11.3		
6	Teller B	6	87	14.5	2.7		
7	Teller C	6	93	15.5	13.1		
8	Teller D	5	110	22	8.5		
9							
10							
11	ANOVA						
12	*Source of Variation*	*SS*	*df*	*MS*	*F*	*P-value*	*F crit*
13	Between Groups	255.6182	3	85.20606	9.694748	0.000498	3.159911
14	Within Groups	158.2	18	8.788889			
15							
16	Total	413.8182	21				

Figure 12.4 The output resulting from the one-way ANOVA tool in Excel's Data Analysis ToolPak.

The first table again gives a few summary statistics of each group separately. For example, these averages (means) do look to be a good bit different. Is this difference statistically significant? The answer is in the second table, the ANOVA table. The 2nd column of this table shows the SSB and SSW, the 3rd column shows the degrees of freedom for the numerator and denominator, the 4th column shows the MSB and MSW, the 5th column shows the value of the test statistic, F, and finally, the 6th column shows the p-value. This is what you compare to α. (The last value in the table is the critical value of F for $\alpha = .05$, against which you can check the test statistic as well.) In this test, the p-value is quite small – definitely less than $\alpha = 0.05$. In fact, we would reject the null hypothesis even at the highly statistically significant level of 0.01. So we conclude that the mean number of customers served per hour by each of the four tellers is not the same. In other words, at least one of the four means is different from the other three.

Exercises

12.1   Use Excel's FDIST function to find the area in the right tail (the p-value) if the test statistic is $F = 1.09$ and $df = (2,12)$.

12.2   Use Excel's FDIST function to find the area in the right tail (the p-value) if the test statistic is $F = 9.69$ and $df = (3,18)$.

12.3   Use Excel's FINV function to find the critical value of F if $\alpha = 0.01$ and $df = (2,12)$.

12.4   Use Excel's FINV function to find the critical value of F if $\alpha = 0.05$ and $df = (3,18)$.

12.5   A dietitian wanted to test three different diets to find out if the mean weight loss for each of these diets is the same. She randomly selected 21 overweight persons, randomly divided them into three groups, and put each group on one of the three diets. The following table records the weights (in pounds) lost by these persons after being on these diets for two months.

Diet I	Diet II	Diet III
15	11	9
8	16	17
17	9	11
7	16	8
26	24	15
12	20	6
8	19	14

Use the one-way ANOVA tool in Excel's Data Analysis ToolPak in order to test the null hypothesis that the mean weights lost by all persons on each of the three diets are the same, using a significance level of 2.5%.

12.6 A consumer agency wanted to investigate if four insurance companies differed with regard to the premiums they charge for auto insurance. The agency randomly selected a few auto drivers who were insured by each of these four companies and had similar driving records, autos, and insurance policies. The following table gives the premiums paid per month by these drivers insured with these four insurance companies.

Company A	Company B	Company C	Company D
75	59	65	76
83	75	70	60
68	100	97	52
52		90	58
		73	

Use the one-way ANOVA tool in Excel's Data Analysis ToolPak in order to test the null hypothesis that the mean auto insurance premiums paid per month by all drivers insured by each of these four companies are the same, using a significance level of 1%.

CHAPTER 13

Simple Linear Regression

CHAPTER OUTLINE

13.1 Simple Linear Regression Analysis
13.2 Linear Correlation

13.3 Hypothesis Testing about the
      Linear Correlation Coefficient

13.1 SIMPLE LINEAR REGRESSION ANALYSIS

For simple linear regression analysis, we will first use Excel's Chart Wizard to generate a scatter diagram for the paired data to be analyzed.

Example 13-1 Suppose a sample of seven households from a low- to moderate-income neighborhood is taken and information on their incomes and food expenditures for the past month is collected. Suppose this information (in hundreds of dollars) is given in the following table. Generate a scatter diagram in Excel.

Income (hundreds)	Food Expenditure (hundreds of dollars)
$35	9
49	15
21	7
39	11
15	5
28	8
25	9

Solution: Type the table into a blank Excel worksheet. Highlight the table and click on the Chart Wizard icon: 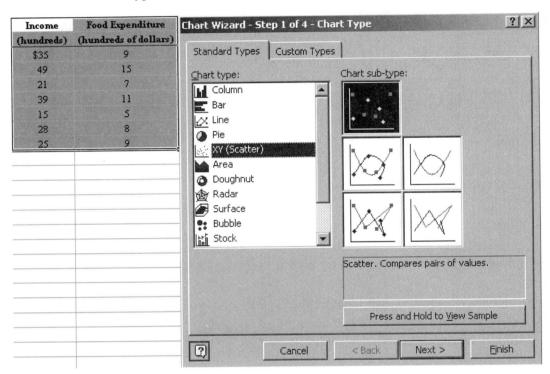. Select the **XY (Scatter)** Chart type and the first option under the Chart sub-type. Then click on **Next.**

Income (hundreds)	Food Expenditure (hundreds of dollars)
$35	9
49	15
21	7
39	11
15	5
28	8
25	9

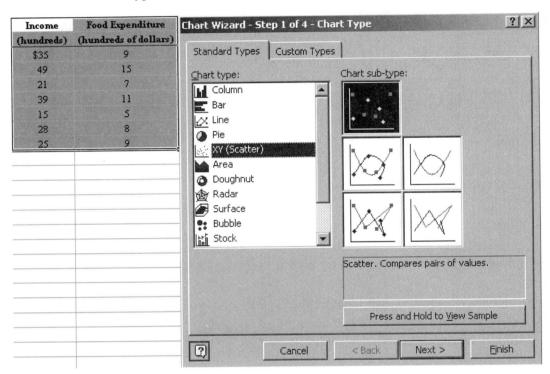

Figure 13.1 Step 1 for generating a scatter diagram.

In Step 2, simply make sure that the data you highlighted is being used to plot the points. You should be able to click on **Next** again with no adjustments.

Step 3 is where you can edit the chart title and the titles for the axes. Also click on the **Legend** tab and uncheck the box next to **Show legend** in order to remove the legend. (Use this label for your *y*-axis instead.) Then click on **Finish.** (Click on **Next** if you want to have the option of creating a new worksheet consisting only of this chart. Otherwise it will automatically place it on the same worksheet with your data.) Resize it (by clicking on it and dragging a corner) or move it (by clicking on it and dragging it from somewhere in the middle) as desired.

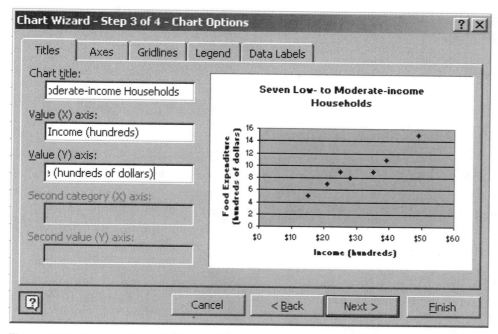

Figure 13.2 Enter and/or edit titles for the chart and axes, and remove the legend in Step 3.

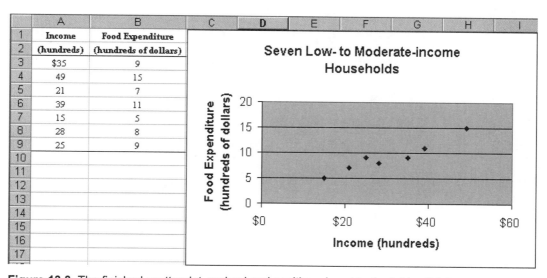

Figure 13.3 The finished scatterplot, resized and positioned next to the table of data.

From the scatterplot, you can see whether it *looks* like there is a linear relationship between the two variables. If so, then we are interested in the least squares regression line for the data. There is an option under the **Chart** menu in Excel whereby you can add this line and its equation to the chart. The **Chart** menu only shows up (along the top, in between **Tools** and **Window**) *when you have clicked on the chart.* (It replaces the **Data** menu which is there otherwise.)

Example 13-2 Find the least squares regression line for the data on incomes and food expenditures for the seven households given in the table in Example 13-1. (As in the scatter diagram, use income as the independent variable and food expenditure as the dependent variable.)

Solution: Click on the scatter diagram generated previously. Go to the **Chart** menu and select **Add Trendline.** Make sure that the **Linear** Regression type is highlighted. Click on the **Options** tab and check the checkbox next to **Display equation on chart.** Leave everything else alone. Then click on **OK.** (If you can't read the equation on the chart because of the gridlines or points, you can click on it and move it to a space where it is more legible.)

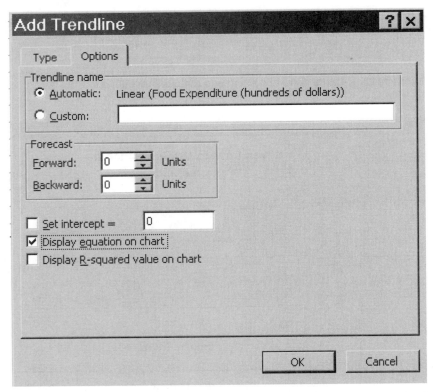

Figure 13.4 Click on the **Options** tab in the **Add Trendline** window and check the box next to **Display equation on chart** in order to have it display the regression line and its equation.

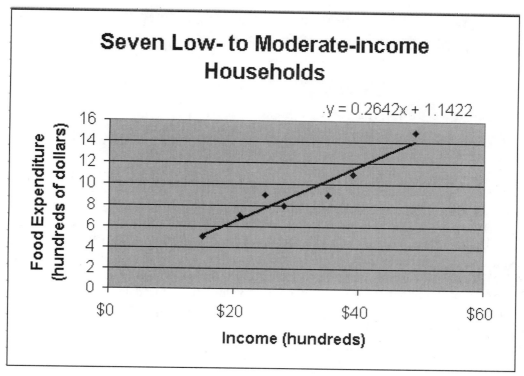

Figure 13.5 The regression line and its equation can be included on the scatter diagram.

You can also have Excel calculate the slope and y-intercept of the equation separately with the SLOPE and INTERCEPT functions. These functions require two array inputs:
1) the array of y-values (the dependent variable) and
2) the array of x-values (the independent variable.

Make sure that you enter them in this reverse alphabetical order!

Example 13-3 Use Excel's SLOPE and INTERCEPT functions in order to find the slope and y-intercept (and then the equation) for the least squares regression line for the data on incomes and food expenditures for the seven households given in the table in Example 13-1. (As before, use income as the independent variable and food expenditure as the dependent variable.)

Solution: Click on an empty cell somewhere near the table of data entered previously, and insert the SLOPE function. (Click on the f_x icon and select it from the **Statistical** category, or just type "=SLOPE(" without the quotes.) For the first input, highlight the *second* column of values from the data table (but not the label). For the second input, highlight the *first* column of values from the data table. Then click on **OK** or hit ENTER. You should see the value of 0.264171 appear. This gives you the slope, which is the coefficient of x in the equation of the regression line.

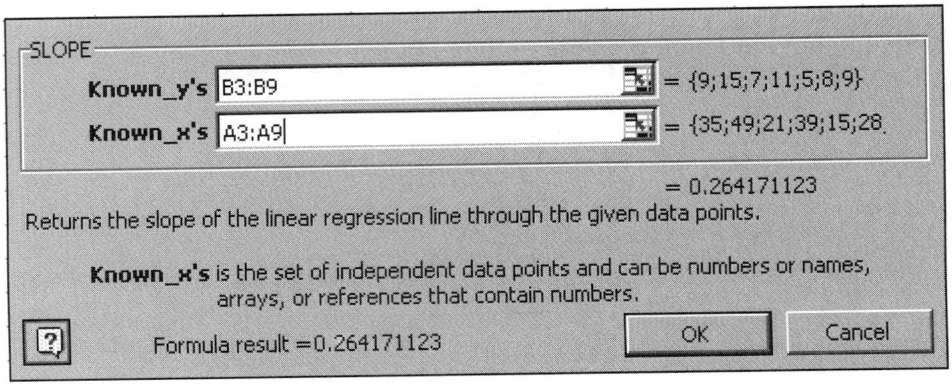

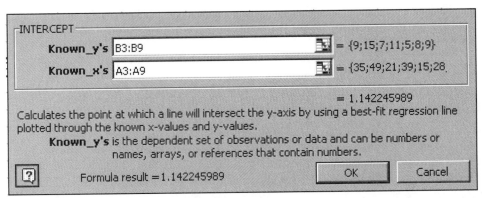

Figure 13.6 Using Excel's SLOPE function to find the slope of the regression line.

Again click on an empty cell somewhere near the table of data entered previously, and insert the INTERCEPT function. (Click on the f_x icon and select it from the **Statistical** category, or just type "=INTERCEPT(" without the quotes.) Again, for the first input, highlight the *second* column of values from the data table (but not the label) and for the second input, highlight the *first* column of values from the data table. Then click on **OK** or hit ENTER. You should see the value of 1.142246 appear. This is the y-intercept, which is the constant in the equation of the regression line.

The equation of the regression line, then, is $y = 0.264171x + 1.142246$.

Figure 13.7 Using Excel's INTERCEPT function to find the y-intercept of the regression line.

Having these individual values for the slope and y-intercept is useful, then, for applying the regression model in order to make predictions.

Example 13-4 Suppose we randomly select a household whose monthly income is $3500. Use the equation of the regression line found in Examples 13-2 and 13-3 in order to predict the value of the food expenditure for this household.

Solution: We are given a value of $x = 35$, since the monthly income values used in Example 13-2 are in hundreds. To calculate the predicted value of y, we need to plug the value of $x = 35$ into the equation $y = 0.264171x + 1.142246$. In other words, we need to multiply this value of x by the slope and add the y-intercept. If you have the values of the slope and y-intercept in separate cells in an Excel spreadsheet, this is very easy. Simply click on an empty cell, type "=" and click on the cell containing the slope of 0.264171, then type "*35+" and click on the cell containing the y-intercept value of 1.142246, and hit ENTER. You should see the value of 10.38824 appear. This corresponds to a prediction of $1038.84 for the food expenditure for this family, since y, also, is in hundreds of dollars.

There is another option in Excel for predicting a value of y when given a value of x. This regression equation is actually built into another Excel function, called FORECAST. This function requires three inputs:
1) the value of x,
2) the array of known y's, and
3) the array of known x's.
The resulting value is the same as what was found above.

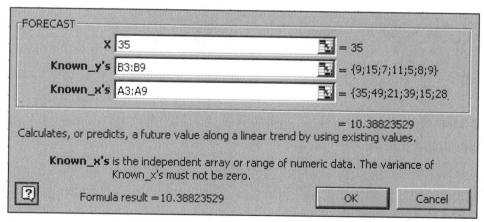

Figure 13.8 Using Excel's FORECAST function in order to predict a value of y for a given value of $x = 35$ for the data as shown in Figure 13.3.

13.2 LINEAR CORRELATION

Of course, a linear regression model is not appropriate when the relationship between two variables is not linear. In this section, we will discuss how to use Excel to find a measure of the strength of the linear association between two variables, the sample correlation coefficient, r.

The function CORREL in Excel will calculate this correlation coefficient. Its two inputs are the same as the SLOPE and INTERCEPT functions: the arrays of the two variables, x and y. Unlike the SLOPE and INTERCEPT functions, though, it is not necessary to enter the y's first, since it comes out the same regardless of which variable is entered first! For consistency, though, we will still enter it this way.

Example 13-5 Calculate the correlation coefficient for the data on incomes and food expenditures for the seven households given in the table in Example 13-1.

Solution: Click on an empty cell near the table of data entered into an Excel spreadsheet and insert the CORREL function. (Click on the f_x icon and select it from the **Statistical** category, or just type "=CORREL(" without the quotes.) For the first input, highlight one of the columns of values from the data table (but not the label). For the second input, highlight the other column of values from the data table. (Though it doesn't matter for this function, in order to be consistent with the other linear regression functions, we will highlight the second column first and the first column second. See Figure 13.9.) Then click on **OK** or hit ENTER. You should see the value of 0.958654 appear. This is the sample correlation coefficient for this data, which shows that there is a strong positive linear correlation. The linear regression model is, indeed, appropriate for this data!

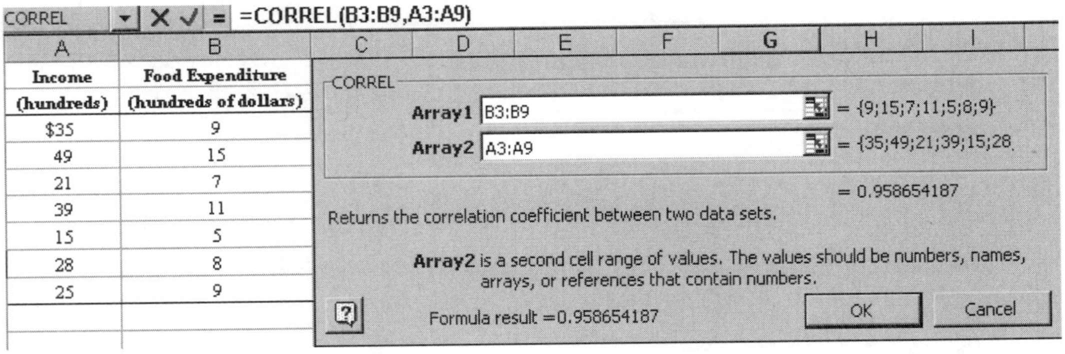

Figure 13.9 Using Excel's CORREL function in order to calculate the sample correlation coefficient, r, for a given set of paired data.

13.3 HYPOTHESIS TESTING ABOUT THE LINEAR CORRELATION COEFFICIENT

In order to determine whether the value of the sample correlation coefficient, r, is statistically significant, we need to conduct a hypothesis test about the population correlation coefficient, ρ, being different from zero (which would indicate no linear correlation).

We will use Excel's TINV function, as in Section 8.2, in order to calculate the critical value, since the t distribution can be used when both variables are normally distributed. Recall that this function requires two inputs:

1) the level of significance, α, for a two-tailed test.
2) the degrees of freedom, which is $n - 2$ in this case.

The test statistic, t, will be calculated via the formula: $t = r \sqrt{\dfrac{n-2}{1-r^2}}$.

The p-value can then be found, as in Section 9.3, with the TDIST function. Recall that this function requires three inputs:

1) the test statistic, t,
2) the degrees of freedom, $n - 2$, and
3) whether the test is 1-tailed or 2-tailed (enter 1 or 2).

Example 13-5 Using the 1% level of significance and the data from Example 13-1, test whether the linear correlation coefficient between incomes and food expenditures is positive. Assume that the populations of both variables are normally distributed.

Solution: From Examples 13-1 and 13-5, we have $n = 7$ and $r = 0.958654$. The claim is that $\rho > 0$, which is the alternate hypothesis. This is a right-tailed test. We have $\alpha = 0.01$.

Click in an empty cell in an Excel spreadsheet and insert the TINV function. For the first input, enter 2*0.01 or 0.02, since this would be the corresponding value of α for a 2-tailed test. For the second input, enter $7 - 2$ or 5. Hit ENTER, and you should see the value of 3.6493 appear. This is the critical value corresponding to the level of significance that we are using.

For the test statistic, we will have Excel calculate the formula given above. Make sure you have the value of $r = 0.958654$ entered (or calculated) in some cell that you can refer to. Click on an empty cell nearby, type "=" and then click on the cell containing the sample correlation coefficient, r. Then type "*SQRT(5/(1-" and click on the cell containing r again. Then type "^2))" and hit ENTER. You should see the value of 7.532729 appear. This is the test statistic. Note that this is greater than the critical value (it is in the rejection region), so we reject the null hypothesis and conclude that there is a positive linear relationship between incomes and food expenditures.

The p-value for this test can then be calculated by inserting the TDIST function with inputs 7.532729, 5, and 1. This produces a value of 0.000326, which also indicates the strong statistical significance of the sample correlation coefficient.

Exercises

13.1 The following table gives the total 2002 payroll (rounded to the nearest million dollars) and the percentage of games won during the 2002 season by each of the National League baseball teams.

Team	Total Payroll (millions)	Percentage of Games Won
Arizona Diamondbacks	$103	60.5
Atlanta Braves	93	63.1
Chicago Cubs	76	41.4
Cincinnati Reds	45	48.1
Colorado Rockies	57	45.1
Florida Marlins	42	48.8
Houston Astros	63	51.9
Los Angeles Dodgers	95	56.8
Milwaukee Brewers	50	34.6
Montreal Expos	39	51.2
New York Mets	95	46.6
Philadelphia Phillies	58	49.7
Pittsburgh Pirates	42	44.7
St. Louis Cardinals	75	59.9
San Diego Padres	41	40.7
San Francisco Giants	78	59.0

a. Have Excel generate a scatter diagram for these data. Does the scatter diagram exhibit a linear relationship between ages and prices of cars?

b. Use Excel's SLOPE and INTERCEPT functions to find the equation of the regression line (with price as the dependent variable and age as the independent variable).

c. Add the regression line and its equation to the scatter diagram from part a.

d. Use Excel's CORREL function in order to calculate the sample correlation coefficient, r.

e. Use the equation of the regression line or Excel's FORECAST function to predict the percentage of games won for a team with a total payroll of $55 million.

13.2 An auto manufacturing company wanted to investigate how the price of one of its car models depreciates with age. The research department at the company took a sample of eight cars of this model and collected the following information on the ages (in years) and prices (in hundreds of dollars) of these cars.

Age	8	3	6	9	2	5	6	3
Price	18	94	50	21	145	42	36	99

a. Have Excel generate a scatter diagram for these data. Does the scatter diagram exhibit a linear relationship between ages and prices of cars?

b. Use Excel's SLOPE and INTERCEPT functions to find the equation of the regression line (with price as the dependent variable and age as the independent variable).

c. Add the regression line and its equation to the scatter diagram from part a.

d. Use Excel's CORREL function in order to calculate the sample correlation coefficient, r.

e. Use the equation of the regression line or Excel's FORECAST function to predict the price of a 7-year-old car of this model.

f. Use the equation of the regression line or Excel's FORECAST function to estimate the price of a 18-year-old car of this model. Is this appropriate?

13.3 Using the 5% level of significance and the data from Exercise 13.1, test whether the linear correlation coefficient between payroll and percentage of games won is positive. Assume that the populations of both variables are normally distributed.

13.4 Using the 1% level of significance and the data from Exercise 13.2, test whether the linear correlation coefficient between age and price is negative. Assume that the populations of both variables are normally distributed.

CHAPTER 14

Nonparametric Methods

CHAPTER OUTLINE

14.1 THE SIGN TEST

The sign test is one of the few nonparametric tests for which Excel is advantageous, so this is the one we will focus on in this chapter. The sign test can be used for tests about preferences about two categories or for tests about medians. All of these tests involve using the binomial distribution with $p = 0.5$. We will show how to have Excel calculate critical values (either in the small *or* the large sample case) with the built-in function CRITBINOM. This function requires three inputs:

1) the number of trials or sample size, $n =$ **Trials**
2) the probability, $p =$ **Probability_s**
3) the cumulative probability = **Alpha**.

The cumulative probability will not always be the given value of α, though. It will be, if it's a left-tailed test. If it's a right-tailed test, then this cumulative probability needs to be $1 - \alpha$. And if it's a two-tailed test, we'll need to apply this function twice, once with $\alpha/2$ and once with $1 - \alpha/2$, to get *two* critical values. Note that these critical values will be the boundaries for the *nonrejection region*, not the rejection region. This is important, since we're dealing with discrete values.

Examples 14-1 and 14-2 show the use of this function for sign tests about categorical data, Examples 14-3 and 14-4 show the use of this function for sign tests about the median of a single population, and Example 14-5 shows its use for a sign test about the median of the differences in paired data. Note that the same procedure can be used both for small *and* large samples, since the binomial distribution is built into Excel even for large *n*. (It is not necessary, therefore, to use the normal approximation for large samples.)

14.2 TESTS ABOUT CATEGORICAL DATA

Example 14-1   The Top Taste Water Company produces and distributes Top Taste bottled water. The company wants to determine whether customers have a higher preference for its bottled water than for its main competitor, Spring Hill bottled water. The Top Taste Water Company hired a statistician to conduct this study. The statistician selected a random sample of 10 people and asked each of them to taste one sample of each of the two brands of water. The customers did not know the brand of each water sample. Also, the order in which each person tasted the two brands of water was determined randomly. Each person was asked to indicate which of the two samples of water he or she preferred. The following table shows the preferences of these 10 individuals.

Person	Brand Preferred
1	Spring Hill
2	Top Taste
3	Top Taste
4	Neither
5	Top Taste
6	Spring Hill
7	Spring Hill
8	Top Taste
9	Top Taste
10	Top Taste

Based on these results, can the statistician conclude that people prefer one brand of bottled water over the other? Use the significance level of 5%.

Solution: The null hypothesis is that the proportion of people who prefer one brand is the same as the proportion of people who prefer the other, or $p = 0.5$, where we can let p be the proportion of all people who prefer Top Taste bottled water. So this is a two-tailed test, and there are two critical values. We have that $n = 9$, since one of the 10 sampled had no preference. And we're using $\alpha = 0.05$. The boundaries for the *non*rejection region, which we will refer to as the critical values for this type of test, are then found with the following functions in Excel:

=CRITBINOM(9, 0.5, 0.05/2) for the lower one, which is 2, and
=CRITBINOM(9, 0.5, 1-0.05/2) for the upper one, which is 7.

As usual, these functions can simply be typed into blank cells in an Excel spreadsheet (with commas between the inputs), or they can be inserted by clicking on the f_x icon and then filling in the text boxes (see Figure 14.1). Either way, the calculated values appear when the ENTER key is pressed.

Since the observed value of x is 6 (the number of people in the sample that preferred Top Taste bottled water), which is in the interval from 2 to 7, the *non*rejection region, the null hypothesis cannot be rejected. We conclude that our sample does not indicate that people prefer either of these two brands of water.

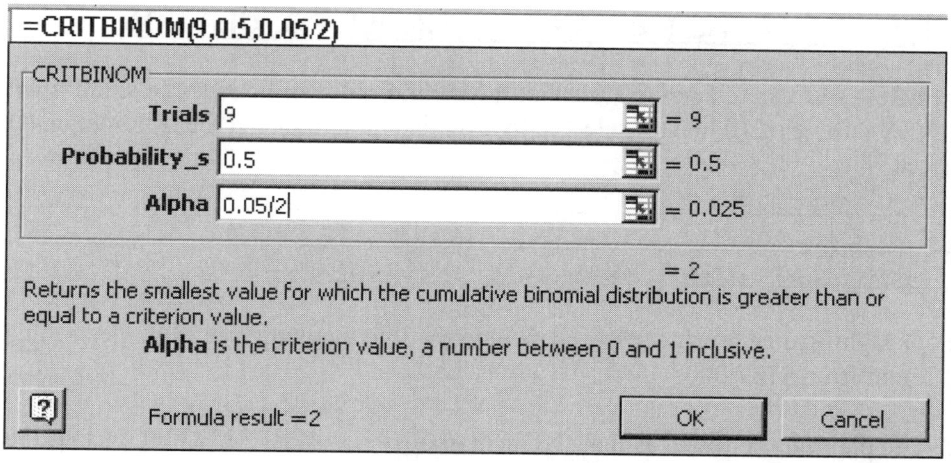

Figure 14.1 Using Excel's CRITBINOM function to find a critical value (which in this case is a boundary for the *non*rejection region) in a two-tailed Sign Test.

Example 14-2 A developer is interested in building a shopping mall adjacent to a residential area. Before granting or denying permission to build such a mall, the town council took a random sample of 75 adults from adjacent areas and asked them whether they favor or oppose construction of this mall. Of these 75 adults, 40 opposed construction of the mall, 30 favored it, and 5 had no opinion. Can you conclude that the number of adults in this area who oppose construction of the mall is higher than the number who favor it? Use $\alpha = .01$.

Solution: The claim to be tested is that $p > 0.5$, where p is the proportion of all adults who oppose construction of the mall. So this is a right-tailed test. Throwing out the 5 adults in the sample with no opinion, we have $n = 70$. And we're using $\alpha = 0.01$.

Since we still have Excel's CRITBINOM function, even for large samples, we don't need to use a normal distribution as an approximation of the binomial distribution.

We can get our critical value, or boundary for the nonrejection region, with the following:

=CRITBINOM(70, 0.5, 1-0.01), which returns a value of 45. Since this is a right-tailed test, this is the *upper* limit of the *non*rejection region. (The lower limit of the right-tailed rejection region is then 46.)

The observed value of x is 40 (the number from the sample who opposed construction of the mall). Since this is less than or equal to 45, it is in the nonrejection region, and we do not reject the null hypothesis. We conclude that the number of adults who oppose the construction of the mall is not necessarily any higher than the number who favor it.

14.3 TESTS ABOUT THE MEDIAN OF A SINGLE POPULATION

Example 14-3 A real estate agent claims that the median price of homes in a small town is $137,000. A sample of 10 houses selected by a statistician produced the following data on their prices.

Home	1	2	3	4	5	6	7	8	9	10
Price ($)	147500	123600	139000	168200	129450	132400	156400	188210	198425	215300

Using the 5% significance level, can you conclude that the median price of homes in this town is different from $137,000?

Solution: The claim to be tested is that the median does not equal $137,000, or that the proportion, p, of prices above $137,000 is not 0.5. So again, we can use the CRITBINOM function with $p = 0.5$ to get our critical values for this two-tailed test. The observed value of x is the number of houses whose prices are above $137,000. (This is referred to as the number of "plus signs" in the terminology of a "sign" test.) We can use Excel's COUNTIF function to determine this count.

Type the table above into a blank Excel worksheet. Click on an empty cell and insert the COUNTIF function. For the **Range** text box, highlight the cells containing the prices. Click in or TAB to the **Criteria** text box and type ">137000" (with *or* without the quotes), and hit ENTER or click on **OK**. You should see the count of 7 appear. This is especially useful when dealing with large samples where it's very difficult to count by hand!

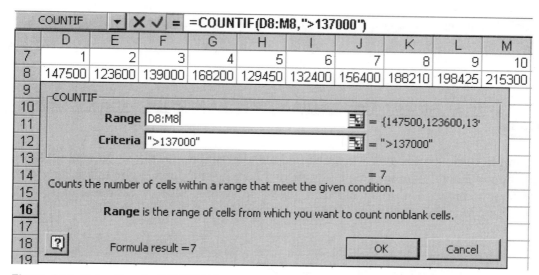

Figure 14.2 Using Excel's COUNTIF function to determine the number of values above the hypothesized median (i.e. the number of "plus signs").

For the boundaries of the nonrejection region, use the CRITBINOM function with the inputs shown below, since $n = 10$ and $\alpha = 0.05$:

=CRITBINOM(10, 0.5, 0.05/2), which returns a value of 2, and

=CRITBINOM(10, 0.5, 1-0.05/2), which returns a value of 8.

Since the observed value is 7, this is in the *non*rejection region, and we conclude that the median price of homes in this town is not any different from $137,000.

Example 14-4 A long-distance phone company believes that the median phone bill (for long distance calls) is at least $70 for all the families in New Haven, Connecticut. A random sample of 90 families selected from New Haven showed that the phone bills of 51 of them were less than $70 and those of 38 of them were more than $70, and 1 family had a phone bill of exactly $70. Using the 1% significance level, can you conclude that the company's claim is true?

Solution: The claim is that the median is at least $70, or that the proportion, p, of families with phone bills above $70 is at least 0.5. (If a larger proportion were above $70, then the median would be higher.) The alternate hypothesis is $p < 0.5$, so this is a left-tailed test. The observed value is $x = 38$ (the number of families from the sample whose bills were more than $70), the sample size is $n = 89$ (throwing out the family with the phone bill of exactly $70), and $\alpha = 0.01$. The critical value, or lower limit of the *non*rejection region, can then be calculated with the following in Excel:

=CRITBINOM(89, 0.5, 0.01), which returns a value of 34.

The upper limit of the rejection region is then 33. Since 38 is greater than or equal to 34, it is in the nonrejection region, and we fail to reject the null hypothesis. Hence, we conclude that the company's claim that the median phone bill is at least $70 seems to be true.

14.4 A TEST ABOUT THE MEDIAN DIFFERENCE BETWEEN PAIRED DATA

Example 14-5 A researcher wanted to find the effects of a special diet on systolic blood pressure in adults. She selected a sample of 12 adults and put them on this dietary plan for three months. The following table gives the systolic blood pressure of each adult before and after the completion of the plan.

Before	210	185	215	198	187	225	234	217	212	191	226	238
After	196	192	204	193	181	233	208	211	190	186	218	236

Using the 2.5% significance level, can we conclude that the dietary plan reduces the median systolic blood pressure of adults?

Solution: The claim to be tested is that the median of the differences in blood pressure (before minus after) is greater than 0, or that the proportion, p, of adults whose blood pressure before the plan was higher than afterwards, is greater than 0.5. So again, we can use the CRITBINOM function with $p = 0.5$ to get our critical value for this right-tailed test. The observed value of x is the number of adults whose blood pressure before the plan was higher than afterwards. (This is the number of "plus signs.") As in Example 14-3, we can use Excel's COUNTIF function to determine this count.

First, type the table above into a blank Excel worksheet. In the cell below each pair, have Excel calculate the difference in each pair (Before minus After). An easy way to do this is to click in the cell below the first pair, type "=" and click on the cell containing the first pair's Before value (or arrow up to it), then type "-" and click on the first pair's After value (or arrow up to it), and then hit ENTER. Once you have this first difference, you can just copy it across to the corresponding cells below the rest of the pairs! Then click on an empty cell and insert the COUNTIF function. For the **Range** text box, highlight the cells containing the differences. Click in or TAB to the **Criteria** text box and type ">0" (with *or* without the quotes), and hit ENTER or click on **OK**. You should see the count of 10 appear. This is the observed value of x for this test.

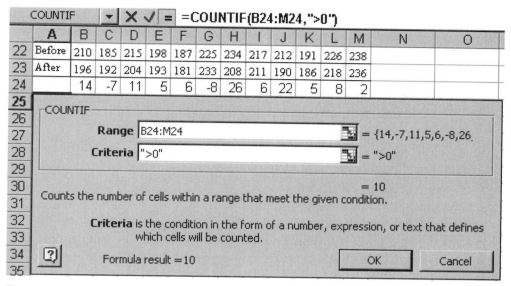

Figure 14.3 Using Excel's COUNTIF function to determine the number of differences above 0 (i.e. the number of "plus signs").

For the upper boundary of the nonrejection region for this right-tailed test, use the CRITBINOM function with the inputs shown below, since $n = 12$ and $\alpha = 0.025$:
=CRITBINOM(12, 0.5, 1-0.025). This returns a critical value of 9.
Since the observed value is 10, which is greater than 9, this is in the *rejection* region, so we reject the null hypothesis and conclude that the dietary plan *does* reduce the median blood pressure of adults.

Exercises

14.1   In the town of Pine Grove, the city water is safe to drink but has a slightly unpleasant taste due to chemical treatment. Some residents prefer to buy bottled water (B), whereas others drink the city water (C). A random sample of 12 residents of the town is taken. The preferences of these 12 individuals are shown here.

B   C   B   C   C   B   C   C   C   C   B   C

At the 5% significance level, can you conclude that the residents of Pine Grove prefer either type of drinking water over the other? Use Excel's CRITBINOM function to determine the critical values.

14.2   Three weeks before an election for state senator, a poll of 200 randomly selected voters shows that 95 voters favor the Republican candidate, 85 favor the

Democratic candidate, and the remaining 20 have no opinion. Using the sign test, can you conclude that voters prefer one candidate over the other? Use $\alpha = .01$ and Excel's CRITBINOM function to determine the critical values.

14.3   The following numbers are the times served (in months) by 35 prison inmates who were released recently.

37	6	20	5	25	30	24	10	12	20
24	8	26	15	13	22	72	80	96	33
84	86	70	40	92	36	28	90	36	32
72	45	38	18	9					

Using $\alpha = .01$, test the hypothesis that the median time served is less than 42 months. Use Excel's COUNTIF function to determine the number of "plus signs" and the CRITBINOM function to determine the critical value.

14.4   Twelve sixth-grade boys who are underweight are put on a special diet for one month. Each boy is weighed before and after the one-month dietary regime. The weights (in pounds) of these boys are recorded here.

Before	65	63	71	60	66	72	78	74	58	59	77	65
After	70	68	75	60	69	70	81	81	66	56	79	71

Can you conclude that this diet increases the median weight of all such boys? Use Excel's COUNTIF function to determine the number of "minus signs" (weight before the diet is less than weight after the diet). Use the 2.5% level of significance and Excel's CRITBINOM function to determine the critical value.

NOTES

NOTES

NOTES

NOTES

NOTES

NOTES